The Changing Academic Library: Operations, Culture and Environment

The Changing Academic Library: Operations, Culture and Environment

Editor

Jina Rowen

scitus
academics

The Changing Academic Library: Operations, Culture and Environment
Edited by **Jina Rowen**

ISBN: 978-1-68117-640-6
Library of Congress Control Number: 2016946500

SCITUS Academics LLC,
616, Corporate Way, Suite 2, 4766,
Valley Cottage, NY 10989

www.scitusacademics.com

Printed in United States of America on Acid Free Paper

Preface

As higher education has evolved through recent decades, college and university libraries also have forged pathways to serve faculty and students more effectively. Academic and research libraries have been early adopters of digital technologies and have provided leadership and training to help remake the academic enterprise. Universities and libraries are operating in a rapidly changing information environment. In this context, libraries have been redefined as facilitators to a world of information far beyond physical collections, providing access to global information through online resources and systems, and also creating their own digital content. Staff and student expectations are developing in line with these changes in the teaching, research and scholarly communication environment. Libraries need to take a strategic role in supporting academics and students within this technologically rich and constantly evolving context. Digital collections and digital scholarship are now mainstream library activities, core to research and teaching. The technology environment will be constantly and rapidly changing, with the expectations and requirements of information users increasing, and with libraries needing to provide an excellent search and discovery experience for their customers. As access to mobile devices becomes ubiquitous, users will expect seamless access to information and services, provided anywhere, anytime.

The Changing Academic Library: Operations, Culture and Environment provides academic administrators and librarians a better understanding of the issues facing the library during this time of change, and the role of the library in the evolving campus of the future. It helps academic librarians plan, implement, and manage changes to the fundamental structure of their organizations.

Table of Contents

1

CHAPTER

Transformational Leadership and Stakeholder Management in Library Change

Dolores Sucozhañay,Lorena Siguenza-Guzman ,Cristian Zhimnay,Dirk Cattrysse, Guido Wyseure,Karel De Witte,Martin Euwema

Faculty of Economics, University of Cuenca, Ecuador, and Research Group on Occupational & Organizational Psychology and Professional Learning, KU Leuven, Belgium

Department of Computer Science, University of Cuenca, Ecuador, and Centre for Industrial Management Traffic & Infrastructure, KU Leuven, Belgium

ACORDES Centre, Faculty of Economics, University of Cuenca, Ecuador Centre for Industrial Management Traffic & Infrastructure, KU Leuven, Belgium

Division of Soil and Water Management, KU Leuven, Belgium

Research Group on Occupational & Organizational Psychology and Professional Learning, KU Leuven, Belgium

Research Group on Occupational & Organizational Psychology and Professional Learning, KU Leuven, Belgium

ABSTRACT

This article aims to analyse the role of library managers as change agent when implementing Library 2.0, using transformational leadership and stakeholder management approaches. To do so, a case study in a Latin American academic library was performed. The experiences acquired for a period of six years were analysed, during which three library managers were involved in managing change. Qualitative data from documents, interviews, and observations were collected, and qualitative analysis methods were used to obtain in-depth understanding of the change process. Results show that lack of transformational leadership and stakeholder management contribute to delayed implementation and limited adoption of innovations. Although library managers recognized the importance of different stakeholders to implement changes, they did not apply systematic and proactive strategies to define and manage them. All in all, library managers should be trained as change agents, with emphasis on transformational leadership and stakeholder management skills.

Keywords: academic libraries, Library 2.0, transformational leadership, stakeholder management

1. INTRODUCTION

The emergence of Internet and Web technologies is not only changing the way people search for information, but also the way information is being published and disseminated (**Jantz, 2012**). Accordingly, to maintain their status as major information providers, libraries are using new technologies for delivering their services (**vanDuinkerken & Mosley, 2011**). Library 2.0 is a revolutionary concept that focuses on user-centred changes, through the use of Web 2.0 tools such as: selective dissemination of information (SDI), forums, social networks, electronic newsletters, enhanced online public access catalogues (e-OPAC), and virtual references. However, this concept is not just about implementing technology; it implies a significant paradigm shift, requiring rethinking and changing traditional library processes and services. Unfortunately, traditional management practices, poor planning activities, lack of management focus, and lack of involvement of important stakeholders may hinder Library 2.0 deployment

(**Munatsi, 2010**). Therefore, to carry out such deep changes, good leadership is key.

Riggs (2001) posits that traditionally, libraries have been over-managed and under-led. This means that although libraries have been well managed, leadership has been pushed aside and overlooked. The words "leadership" and "management" are often erroneously used interchangeably (**Riggs, 2001**). In fact, management and leadership are fundamentally different skills; management is related to performing existing practices in an adequate and efficient way, while leadership is defined as setting, motivating and aligning people with a new direction (**Kotter, 1996**). Therefore, Library 2.0, a philosophy of continuous changes, demands that leadership no longer be ignored and requires library managers or administrators to rethink the way they lead (**Riggs, 2001**). In this article, the term "library manager" is used to refer to a director or the formal position of authority.

The study of library leadership has increased significantly in the last years. Although different qualities of effective leaders have been identified, transformational leadership has been found as more effective in managing libraries coping with complex changes (**Castiglione, 2006**; **Jantz, 2012**; **McGuigan, 2012**). Transformational leadership motivates followers to go beyond their own interests and move closer towards the organizational interests (**Yukl, 2013**).

Library managers have considerable impact on library change; however, different stakeholders, such as university authorities and library staff, also have a strong influence on these changes (**López & Vargas, 2012**; **Solís & López, 2000**; **Stavridis & Tsimpoglou, 2012**). In fact, the internal culture, structure and decision-making of the academic libraries are also influenced by the 'mother' organizational culture (**vanDuinkerken & Mosley, 2011**). Nevertheless, most library leadership literature focuses on the leader-follower relationship (**Tam & Robertson, 2002**; **Williamson, 2008**), which limits the understanding of the relationships among library managers and other university actors.

The aim of this article, therefore, is to highlight the role of library managers in implementing Library 2.0, by focusing on their leadership style, and stakeholder management practices. The combination of these two essential tasks has not, to our knowledge, been studied before in this context. To do so, the experiences acquired during six years of attempted change in an academic library in Latin America have been analysed. This academic library was selected as it belongs to one of the most prestigious universities, and is considered one of the most modern and largest libraries of its country. The subsequent section recalls briefly the main theoretical ideas concerning leadership and stakeholder management in change (§2), followed by a description of the procedures used in this study (§3), leading to the delineation of its findings (§4), discussion (§5) and conclusions (§6).

2. LITERATURE REVIEW

There is a plethora of transformational leadership and stakeholder management literature from business arenas. However, according to **vanDuinkerken and Mosley (2011)**, it is still difficult to understand how to apply these insights to the library environment and, more precisely, to academic libraries.

2.1. Transformational leadership in academic library change

Bass (1990) suggests that leaders who exhibit transformational behaviours can influence followers to go beyond their own interests and move towards the organizational interests. In fact, during changes in academic libraries, transformational leadership has been found in several case studies to be more effective than other leadership styles, such as transactional and laissez-faire leadership (**Castiglione, 2006**; **Jantz, 2012**; **McGuigan, 2012**).

Transformational leaders achieve extraordinary levels of follower trust, admiration, motivation, commitment, loyalty and performance (**Kouzes & Posner, 2012**; **Yukl, 2013**), by means of four dimensions:

1. Idealized influence
2. Inspirational motivation
3. Intellectual stimulation and
4. Individualized consideration

Bass and Riggio (2006), and **Kouzes and Posner (2012)** describe these four dimensions more in detail. For instance, idealized influence is defined as serving as role models for perseverance and self-sacrifice; leaders are admired, respected and trusted. Inspirational motivation provides an appealing vision, and stipulates how the team can achieve the desired future. Intellectual stimulation encourages followers to be innovative and creative by analysing issues from a new perspective. Finally, individualized consideration means to pay attention to each individual's needs and desires, including coaching, supporting and a two-way exchange communication. **Jantz (2012)** in his study concludes that transformational styles empower librarians to create a more innovative environment, and **Castiglione (2006)** points out that organizational learning and adaptation are encouraged by transformational behaviours.

Transformational leadership is often explained in relation to transactional leadership. The latter, also known as managerial leadership is mainly defined as an exchange process in which leaders "promise and deliver rewards to employees for carrying out assignments" (**Bass, 1990**, p. 30). According to **Crosby (1996)**, transactional leadership focuses on managing the status quo with emphasis on managing subordinates by standard work, establishing well-defined processes, and managing through accomplishable tasks. Transactional leadership is characterized by two leadership behaviours: active management by exception, and contingent reward. Active management by exception is defined in terms of looking for and dealing with irregularities, mistakes, exceptions and deviations to ensure that standards are met (**Antonakis, Avolio, & Sivasubramaniam, 2003**; **Bass, 1990**). Contingent reward refers to leader behaviours that focus on both the clarification of expectations and task requirements, and offers the recognition of followers when goals are achieved (**Antonakis et al., 2003**; **Bass, 1990**).

Avolio and Bass (1991) added to their initial theory the laissez-faire or passive leadership style, which is associated with laissez-faire and passive management by exception. Laissez-faire – or "no leadership" is the absence of effective leadership, avoiding decision-making, abdicating responsibilities, and not using their authority. This leadership behaviour can be considered active to the extent that leaders choose to avoid taking actions (**Antonakis et al., 2003**). Passive management by exception is characterized by responding passively and reactively to situations and problems systematically. For instance, library managers displaying passive leadership behaviours, expect librarians to solve problems on their own without his/her direct supervision. Moreover, laissez-faire leaders do not provide a vision to be achieved by followers (**Avolio & Bass, 1991**).

Bass (1990) considers that every leader displays each of the three aforementioned styles to some extent, representing the elements of the so-called "Full Range of Leadership Model (FRL)". In this FRL model, leadership effectiveness is ranged from highly active and effective, to highly inactive and ineffective. Transformational leadership is considered highly active and effective with an emphasis on development and change (**Yukl, 2013**). Transactional leadership is situated in the middle; its effectiveness is considered in terms of focusing on irregularities, mistakes, exceptions, and deviations from standards (**Verschueren, 2014**). Laissez-faire or passive leadership style is associated with the most passive and most ineffective leadership. Both styles, transformational and transactional, can be complementary, though change effectiveness is related to greater use of transformational behaviours (**Huang & Liao, 2011**; **Yukl, 2013**).

Yukl (1999) posits that the current emphasis on the relationship between transformational leader and subordinate should be complemented with a more organizational perspective. Implementation of change implies not only influencing subordinates, but also other stakeholders at different levels (**Armenakis & Bedeian, 1999;Cummings & Worley, 2005**; **Govender, Moodley, & Brijball Parumasur, 2005**; **Kotter & Cohen, 2002**; **vanDuinkerken & Mosley, 2011**; **Yukl, 2013**). Thus, as **Kotter and Cohen (2002)** suggest, leaders should ensure that all actors with a stake in the change be identified and involved in the process and implementation of that change. Unfortunately, this seems not to be the case in most academic libraries, as leaders have scarce communication with stakeholders during change processes (**Stephens & Russell, 2004;vanDuinkerken & Mosley, 2011**).

2.2. Stakeholder management in academic library change

The stakeholder management approach offers an interesting frame to understand the need and the way to actively manage stakeholders in change settings. **Freeman (1984)**defines a "stakeholder" as any group or individual who can affect, or is affected by, the achievement of the change's objectives. Stakeholders are only defined in reference to a particular issue (**Bryson, 2004**; **Savage, Nix, Whitehead, & Blair, 1991**). In general, the stakeholder management approach involves four steps: (1) identification and analysis of stakeholders, (2) formulation of strategies, (3) facilitation of the process to implement these strategies, and (4) evaluation of strategy implementation (**Freeman, 1984;Mitchell, Agle, & Wood, 1997**; **Savage et al., 1991**).

Savage et al. (1991) developed a model to apply stakeholder management based on an analysis of the potential for stakeholders to threaten the process, as well as their potential to cooperate. The authors recommend that leaders should assess these two potentials and manage the stakeholders accordingly. They also suggest minimally satisfying the needs of stakeholders with low potential on both dimensions, while maximally satisfying the needs of stakeholders with high potential to cooperate, and high potential to threaten. The fundamental idea is to transform the stakeholder relationship to a more favourable condition (**Bryson, 2004**).

Library managers, as part of their role as change agents, should implement this approach. However, leaders often only identify and describe stakeholders but fail to successfully manage them (**Freeman & McVea, 2001**). Stakeholder management should be seen as a continuous task of analysing and integrating multiple stakeholders and multiple objectives (**Freeman & McVea, 2001**). Furthermore, the criteria used to assess stakeholders are not static but rather dynamic. Levels of power, saliency, coalitions and interest, may change over time, and leaders must pay attention to these changes (**Freeman, 1984**).

Concisely, whereas transformational leadership is directed at the library staff, stakeholder management focuses on other relevant parties. Based on the characteristics of transformational leadership mentioned before, it is likely that library managers with more transformational styles will also deploy more stakeholder management.

3. METHODOLOGY

A single case study concerning a six-year attempt to implement Library 2.0 into a Latin American academic library was used. This allows for a better understanding and interpretation of the events, leadership behaviours, and relationships between the library manager and the different stakeholders during the change process.

Data collection was accomplished through direct observation, analysis of secondary documents, and in-depth semi-structured interviews to assure data triangulation (**Yin, 2009**). This data collection was conducted by the first author over the six year period in order to appropriately cover the whole process and to avoid errors caused by a prospective memory failure. Through sustained observation, deep level understanding of the process was achieved. Forty official and unofficial documents were collected, including change plans, meeting reports, communications, change process reports, library regulations, strategic plans, university regulations, and university newsletters. Comprising the data were fifteen semi-structured interviews with key informants: library managers (3 interviews), librarians (9), library council members (2) and university's top administrators (1). The first round of interviews started in the second year, and continued yearly thereafter. Each interview lasted approximately 35–60 minutes. The interviews were recorded and then transcribed for analysis. The topics explored in the questionnaire, as listed in Appendix 1, were the change process, change´s results, leadership behaviours, leader relationships, stakeholder management, and limitations for change implementations.

The data analysis was carried out in several steps, mainly using qualitative software (ATLAS.ti). Firstly, data were studied and categorized meticulously in order to identify and describe the change periods and results. Codes and subcodes were used to organize the data in each period. For instance, the code "CP_Time line" was used to code the milestones of the process, and the codes "CP_First Period", "CP_Second Period", "CP_Third Period" and "CP_Fourth Period" were used to organize the events in each period. In addition, the codes "CP_Planning", "CP_Objectives", "CP_Implementation", "CP_Results" and "CP_Evaluation" were utilized to code and analyze the steps followed to implement the change, as well as the outcomes achieved in each phase.

(See details in Appendix 2.) Secondly, a priority list of codes, based on our previous literature review (§ 2) and emergent ideas from the data, was used for coding. A detailed list of codes is given in Appendix 2. To ensure accuracy of coding, data were read and independently coded by two of the authors, and then were jointly compared in order to consolidate the codification. Quotations, which had been coded differently, were discussed in detail until consensus was achieved. Thirdly, matrices were developed in order to summarize and clarify codes and levels of abstraction. For more information about the use of matrices see **Miles and Huberman (1994)** and **Swanborn (2010)**. An example of a summarized matrix is given in Appendix 3. Finally, exploring data tools of Atlas.ti and matrices were used to generate findings. These emerging findings were discussed by the authors, and continuously compared and contrasted with literature on transformational leadership and stakeholder management in order to identify its emergent theoretical contribution to organizational change (**Jorgensen, 1989**).

Research quality checks were accomplished using three different procedures. First, independent coding was used. Second, the interview findings were triangulated with findings from the secondary documents and observations, avoiding dependence on the interviewees' subjective views. Third, this study employed a member checking process.

4. THE ACADEMIC LIBRARY CHANGE PROCESS

The main objective of the change process was to achieve the transition from a traditional library to a Library 2.0. The key drivers for starting this change were both the pressures of the changing context, and the implementation of an Institutional Change Project (ICP) at the university level with financial support from an external institution. The

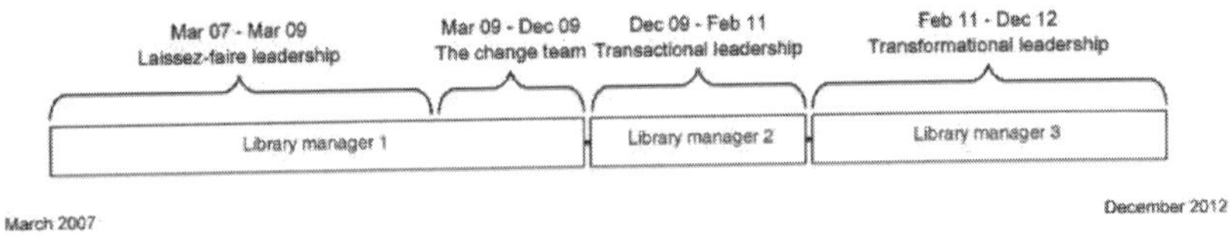

Figure 1. Overview of the periods of the library change process

Table 1 Summary of findings.

Period/ Library managers	**Transformational Leadership***	**Stakeholder management ****	**Changes implemented**
Laissez-faire leadership Library manager 1	No transformational behaviours exhibit.	Hardly stakeholder management practices performed.	Technological changes****
The change team Library manager 1 + Change Team***	No transformational behaviours exhibit.	Library Council integration, only *structured attempt* to involve stakeholders.	Technological changes****
Transactional leadership Library manager 2	No transformational behaviours exhibit.	Stakeholder management was not systematic, nor proactive. Good relationships with specific stakeholders.	Technological changes
Transformational leadership Library manager 3	Transformational behaviour exhibit.	Stakeholder management was not systematic, nor proactive. Good relationships with specific stakeholders.	Technological and organizational changes

* Transformational leadership behaviours: individualized consideration, inspirational motivation, intellectual stimulation, idealized influence.
** Stakeholder management practices: identification, analysis, formulation of strategies; facilitation of the process to implement these strategies and evaluation of strategy implementation.
*** The results are referred to LCT. Results for the library manager are similar to the first period.
**** Technological changes were in charge of the ICT responsible.

ICP´s leader was the Vice Rector, responsible for all academic affairs including academic programs, library services and students' well-being.

The six years of change were divided in four key periods related to important milestones and shifts in library manager (see Figure 1).

Subsequently, the findings are presented for each period as follows. First, the main events and results are described. Then, the role played by library managers as leaders of change and how they dealt with different stakeholders are analysed (see Table 1). Finally, a comparison of the findings regarding leadership and stakeholder management is provided.

4.1. First period: "Laissez-faire" leadership (March, 2007–March, 2009)

The change began in March 2007 when the ICP at the university level started to set up the library change as one of its sub-projects. A diagnostic of the library, on initiative of the ICP, was developed by an external group of international library experts, including a young and a senior chief librarian. Several recommendations were identified and communicated with the university's top administrators and librarians. Although several meetings and informal conversations took place among library staff and the ICP, the planning of change was not clearly and explicitly formulated. After almost two years, the interviewers recognized that the change had not been successfully implemented. This conclusion was corroborated by reports and observations. Organizational changes, such as a new library regulation and a new departmental structure, were not executed. A number of technological improvements were implemented by the ICT responsible, including: the installation of new computers, servers and storage equipment; the creation and implementation of a new Website; and higher-speed access to Internet and Intranet. After several discussions, the need to retake efforts to boost the change process in the library was recognized by top management. See **Sucozhañay et al. (2011)** for a detailed description of this period.

Regarding the leadership style, the library manager who was expected to take the lead in implementing the change, was perceived as playing a rather low-profile role and not exhibiting transformational leadership behaviours. Librarians saw him/her as someone who was highly experienced and trained but who failed to provide them with regular feedback. The library manager's leadership style was characterized as allowing librarians great freedom to propose and execute activities while avoiding getting involved and making decisions. An interesting fact, regarding the leadership on this period, is the little clarity on who was leading the change. For instance, on the one hand, the technological changes were conducted by the ICP and ICT responsible. On the other hand, when the library manager was asked about the leadership, he/she identified the ICP as leader of the change. Likewise, the Vice Rector, as the highest authority of the library, was also identified as responsible for the change. Unfortunately, the vision of change of these different actors was not aligned. The library manager explained that his/her idea of change was oriented to solve librarianship issues, such as cataloguing and processing, while the Vice Rector and ICP in coalition with the ICT responsible, were mainly focused on technological improvements. This disconnection contributed to the isolation of the library manager.

In addition to this lack of alignment, stakeholder management practices were not observed or mentioned in the interviews. The relationship between the library manager and stakeholders was weak; consequently, neither influenced the librarians, nor university's top administrators, and other actors to align forces for change (as shown inFigure 2a).

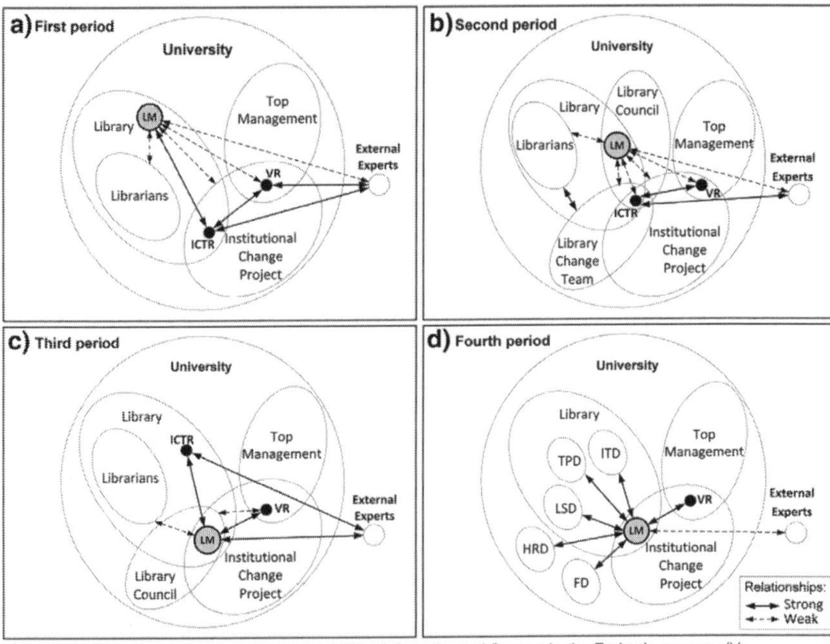

Note. – LM = Library Manager. VR = Vice Rector. ICTR = Information and Communication Technology responsible. HRD = Human Resource Department. FD = Financial Department. ITD = Information Technology Department. TPD = Technical Process Department. LSD = Library Services Department.

Figure 2 Library managers' relationships with key stakeholders during the change attempts.

4.2. Second period: The change team (March, 2009–December, 2009)

After the first attempt to forge the change into the library, a new strategy was planned. Thus, an external expert updated the diagnosis of the library and drew the library change path but now with the support of a local Library Change Team (LCT) and a Library Council. The LCT was formed by an organizational consultant, a thesis student, an internship student of a European university, and the ICT responsible. The Library Council consisted of representatives from academic and library staff, and university top and middle management, formally appointed by the Vice Rector.

The diagnosis showed that the library situation had not changed, except for some technological improvements. After several meetings with the participation of the library staff, LCT, Library Council and the external expert, a general guide for the library change, including vision, mission and five main objectives, was set out and validated by university's top administrators. The objectives were: 1) adapting the organizational structure, 2) optimizing the internal processes, 3) improving the interpersonal relationships, 4) positioning of the library image, and 5) updating the library regulations. The LCT took charge of the implementation process. Important to note is that the library manager was not an active actor of the LCT, due to, as mentioned above, his/her misalignment of the change vision.

During the next six months, the activities were geared to implement this plan. The technological change, led by the new ICT responsible, continued. A new library system and a new cataloguing format were implemented. However, the gap between the implemented technology and ideas, and knowledge and skills of the library staff remained wide. As several librarians explain, many technologies and services were implemented without clearly knowing the theory and philosophy behind them. For example, Facebook services were implemented without a clear idea of its importance as part of Libraries 2.0. More details on this period can be found in **Sucozhañay et al. (2011)**.

Concerning the leadership of this period, no change in the library manager performance was found. The LCT had a prominent role in this period. This team had a direct communication with the Vice Rector and librarians. Nevertheless, transformational behaviours of its members were not found. The limitation of the LCT was its little political power to take decisions, since none of its members had a formal position at the university.

Regarding stakeholder management practices, the relationship between the library manager and the stakeholders remained relatively poor (see Figure 2b). No systematic stakeholder management practices were identified. The establishment of the Library Council could be considered as a "*first structured attempt*" to involve several stakeholders though this initiative was boosted by the ICP. The LCT carried out the establishment of the Library Council. Unfortunately, university's top administrators and the library manager hardly participated in the Library Council, thereby minimizing the potential stakeholder management.

4.3. Third period: The managerial leader (December 2009–January 2011)

In December 2009, the university's top administrators appointed the ICT responsible as library manager. The good relationship and shared understanding of the content of change allowed direct communication between the library manager, ICP members, and the Vice Rector, thereby eliminating previous disconnections. Activities in this period were focused on the implementation of proposals that were developed in the previous period, such as the new library regulation, training of library staff, and dissemination of online services. Despite the emphasis placed on the development of proposals for a new organizational structure, including departmentalization schemes, process optimization, and manual functions, none of these contributions were implemented. Librarians and Library Council members perceived that no big organizational changes were achieved. The technological change continued to be an important element in the library change. As a result, a new online public access catalogue was implemented, and the library Web portal was redesigned to incorporate Library 2.0 tools, such as blogs, social networking, messaging applications and Really Simple Syndication (RSS) feeds.

Regarding leadership behaviours, the majority of librarians perceived the new library manager as more active, communicative, task-oriented, and having a high commitment with the change process. The library manager put emphasis on setting short-term goals, rules and procedures. Despite these favourable characteristics, the new library manager was not perceived by librarians as having enough influence, librarian knowledge, and expertise to motivate and inspire followers. The relationship between the new library manager and librarians tended to be transitory and not based on emotional bonds. Accordingly, transformational behaviours were almost absent. Interestingly, although librarians recognized the need for change, it did not necessarily mean radical change, and they often felt threatened by the new technological innovations promoted by the library manager.

Stakeholder management in this period was not systematic but rather incident-based. For instance, the library manager directly worked together with university's top administrators to achieve specific results such as the final approval of the new library regulation. The Library Council continued having meetings and discussions on how to implement the change and how to get enough resources for it. Additionally, they provided the political support and feedback for the process of approval of the new library regulation. Interestingly, after almost one year, the work intensity of the Library Council decreased due to both few meeting requests and the intermittent attendance of their members. The relationships of key stakeholders in this period are presented in Figure 2c.

The analysis of this period prompts the conclusion that although the new library manager had a good stakeholder relationship, he/she did not display transformational leadership qualities. This situation limited the library manager's ability to influence stakeholders to achieve the library vision.

4.4. Fourth period: The transformational leader (February 2011–December 2012)

At the end of January 2011, the second library manager resigned to accept another position. As a consequence, an experienced librarian was appointed as new library manager by university's top administrators.

In this period, the new library manager was able to implement the new organizational structure that had not yet been implemented, due to low participation and commitment rates of librarians. The new organizational structure consisted of one head and three departmental units: Technical Process, Library Services, and Information Technology, each with their respective coordinator. After its implementation in April 2011, librarians noted that the new structure allowed for the specialization, promotion, teamwork, improvement of coordination, and monitoring of library functions.

With the new structure implemented and running, efforts were made to develop the library strategic plan for 2012–2017. To give support in this task, a new external expert visited the library in May 2011. After a week of work, the library staff and university's top administrators agreed on the strategic lines. Librarians recognized the importance of the plan in clarifying the library direction for the upcoming years. The strategic lines were: 1) organization and co-operation, 2) support of research, 3) support of teaching and learning, 4) digital library, 5) staff capacity building, and 6) quality management.

The emphasis in this period was then put on the implementation of the library strategic plan. Digital services, such as a digital repository and remote access to bibliographic databases, were implemented. Library staff and users were trained in topics related with digital services. Additionally, the library participated actively in the university accreditation and improved its cooperation and networking with other libraries.

The new library manager was recognized by all librarians as leader of the change process and displaying transformational leadership qualities. Attributes, such as communicative, persistence, envisioning future, commitment to goals, valid knowledge and skills, ideas generator, team building among others, were used to describe the library manager's leadership style. Attitudes that, according to the literature, fall into the four transformational behaviours: idealized influence, inspirational motivation, intellectual stimulation, and individualized consideration (**Bass & Riggio, 2006**). The following quotation illustrates this view: "*The library manager is leading the process of change, and has become a leader because of the determination, knowledge and passion about the library*"(Librarian).

Regarding the stakeholders, the library manager recognized the importance of the Library Council because its members are important stakeholders; however, no effort was made to reactivate it. Instead, to implement certain activities, the library manager initiated direct communication with university's top administrators and different university departments and units, such as Human Resources and the Financial Department (see Figure 2d). Although a good relationship with these stakeholders could be seen, neither a systematic nor a proactive management of the stakeholders could be identified. An overview of the findings implemented in all investigated periods, regarding transformational leadership, stakeholder management, and main changes, is given in Table 1.

5. DISCUSSION

The transition from a traditional library to a Library 2.0 in Latin America should be seen as an integral and complex change. As can be seen in this case study, implementing these types of changes requires a good deal of vision, time, and effort (**Jantz, 2012**). Technological and organizational components have to be managed simultaneously;

otherwise, this shift of paradigm must settle for merely automating traditional library services. Accordingly, **Graetz and Smith (2009)** argue that many libraries continue to uphold traditional structures while they try to implement new approaches. In fact, the key to successful implementation of any new model is usually not the technology but human factors (**Williamson, 2008**), and this case study is not the exception. Evidence shows that people are essential to the success of a change process, but they often resist (**Marshall, 2008**). Zimmerman (2006) posits that resistance is a major factor in the failure of change. Therefore, library managers have to develop a participative process that encourages librarians and other stakeholders to work together to achieve a common goal, while preventing significant resistance to change (**De Bruyn & Kruger, 2001**). In this case, change resistance was not highly significant, it was mainly related to the way that the change was led.

The literature on leadership styles and stakeholder management provides insight into the role of library managers in the implementation of change. The current findings reinforce the conclusion that transformational behaviours contribute to positive outcomes during times of change in academic library settings (**Albritton, 1998**). Transformational leadership style stresses the emotional and symbolic aspects of change (**Yukl, 2013**). For instance, leaders usually use emotional support to prompt followers' commitment, and thus reduce resistance. **Stringer (2002)** posits that leadership practices cannot be measured as 'good-bad' issues but rather as 'effective-less effective' issues. In this case, the third library manager, who displayed higher transformational leadership behaviours, was able to inspire, motivate, and facilitate change not only in technological aspects but also in organizational aspects. A transformational leader motivated library staff by means of communicating an appealing vision, demonstrating high levels of trust, persistence and knowledge about the library issues, attention to individualized development, and the ability and willingness to provide intellectual stimulation (**Bass, 1990**, p. 30). The second library manager was recognized as task-oriented and less effective in achieving an integral change. This insight further supports the idea that although transactional leadership style is appropriate in some settings, only transformational leadership can motivate the actions needed to achieve successful change (**Kotter, 1996**).

Additionally, during the whole process, only one of the three library managers was perceived as a transformational leader. This drives our attention to the way library managers were appointed. It was found that no analysis of leadership skills was performed before the appointment of library managers to their positions. This finding is in agreement with **Sukram and Hoskins' (2012)** findings, which show that library managers are often appointed without considering their leadership skills. In this study, it is therefore argued that one "Achilles' heel" of the library change was to assume that appointed library managers would be able to serve as transformational leaders. Traditionally, the selection of library managers is the exclusive responsibility of the university's top administrators. Unfortunately, these authorities lack time and experience to assess the candidate's leadership qualities.

Furthermore, as mentioned in the literature review, academic libraries are usually part of a parent organization, such as a university. This makes library change seen as a phenomenon that occurs in interaction with a wider range of relevant stakeholders (**López & Vargas, 2012**; **Solís & López, 2000**). For instance, in this case study, strategic decisions and resource allocation depended not only on library managers' decisions but also on university top administrators and other stakeholders, such as the finance director and deans' decisions (**Budd, 2005**; **López Segrera, 2008**; **Torres & Schugurensky, 2002**;**vanDuinkerken & Mosley, 2011**). This fact drives the discussion of the stakeholders' value in developing a major change in library settings, which has been corroborated in previous studies, such as **vanDuinkerken and Mosley (2011)**. According to **Turner**

(2007)and **Kotter (1995)**, stakeholders' attitudes, including resistance or willingness to change, depend not only on the content of the change but largely on how the transition is led. Nevertheless, most of the library leadership literature has focused on the leader-subordinate relationship (**Tam & Robertson, 2002**; **Williamson, 2008**).

The library manager, as change leader, has to engage stakeholders with the goals of the change (**Stavridis & Tsimpoglou, 2012**; **vanDuinkerken & Mosley, 2011**). However, as illustrated in Figure 2, findings show that most of the time, library managers have a rather limited and weak relation with stakeholders (**vanDuinkerken & Mosley, 2011**).**Horn (2008)** posits that each stakeholder may require a different approach to be managed. For instance, library managers have to be aware of the strong contribution that university top management can make to achieve radical change. Nevertheless, even though support and commitment to change is expected and offered from university's top administrators, it fluctuates during the change process. Low attention and engagement are sometimes evidenced. Therefore, library managers have to work on building and strengthening their relationship with university top management to increase the chances to succeed. Likewise, university's top administrators, as principals responsible for library development, should take sound decisions to avoid bottlenecks during the implementation of change, such as providing budget and resources.

Stakeholder management practices were barely identified in our case study. Nevertheless, it does not necessarily mean that library managers developed different relationship approaches with their stakeholders. Remarkable is that, although all three library managers recognized the importance of university actors, none of them performed a systematic practice to identify, analyse, and manage these stakeholders. This finding reinforces the reported experience on previous studies that leaders often fail to identify, analyse, and manage stakeholders (**Savage et al., 1991**). From a pragmatic point of view, the importance of stakeholder management must be recognized; managing highly diverse stakeholder groups effectively, is not an easy task as demonstrated in this case study. There were some isolated practices that library managers performed in order to negotiate and engage some stakeholders. For instance, the "Library Council" experience was an interesting attempt to incorporate other stakeholders. Nevertheless, even with a promising start and recognition as part of the organizational structure of the library, the committee stopped its work by the end of the third period. Although this experience would have been an interesting way to integrate stakeholders, it had its limitations. First, the selection of Library Council members was guided by a criterion of "experience in research" but not according to their power, legitimacy, and urgency (**Mitchell, Angle, & Wood, 1997**); three key stakeholder attributes that have received reasonable empirical support to identify and analyse "who really counts" when performing stakeholder mapping and evaluation (**Laplume, Sonpar, & Litz, 2008**; **Mitchell, Angle, & Wood, 1997**). Second, no analysis or management strategies of these actors was generated by leaders. For instance, most of the times, key stakeholders were very busy with the day-to-day management of their own faculties or departments, simply lacking time and energy to invest in collaboration with other parties. Finally, Library Council recognition in the third period was still low, limiting its capacity to influence decision-making processes.

Several methodological proposals on how leaders can manage stakeholders can be found in the literature (**Mitchell, Angle, & Wood, 1997**; **Savage et al., 1991**; **Svendsen, 1998**). However, it is not clear why leaders do not put any of these recommendations into practice. One possible explanation, coming from this study, is that library managers usually see the university top administrators and the administrative and academic directors as individuals to whom they should report, rather than actors to be influenced. Another possible explanation is the lack of time library managers and stakeholders have for such activity. Based on the case study analysis, the general conclusion is that although the importance of stakeholders is recognized by library managers, their management is more reactive than proactive. This prompts the need of library leaders to understand more deeply the stakeholder management approach, which means to recognize the stakeholders, their motivations, and interests. In addition, to increase the likelihood of a successful change, an important activity is the formulation of strategies, and subsequently the facilitation of processes to implement such strategies (**Mitchell, Angle, & Wood, 1997**; **Savage et al., 1991**).

5.1. Practical implications

In general, our findings suggest that library managers, as change leaders, should perform transformational behaviours and actively manage the stakeholders in order to achieve a successful change. This double approach – transformational leadership and stakeholder management – used to analyse the role of library managers, recognizes the complexity of relationships in an academic library change implementation.

Sukram and Hoskins (2012) recommend that library managers be trained on leadership competencies and management of the stakeholders. Nevertheless, managers are rarely trained in these issues (**Holm-Nielsen, Thorn, Brunner, & Balán, 2005**; **vanDuinkerken & Mosley, 2011**). Library managers have to look for transformational leadership skills to encourage organizational change. **Bass (1990)** posits that transformational leadership skills can be learned, and plenty of methodologies to implement stakeholder management can also be found in literature. However, as these concepts have been developed mainly in business arenas, library managers have to adapt them for the academic field. Training in both transformational leadership and stakeholder management will help library managers improve their performance as change leaders.

5.2. Research limitations

A limitation of this study is that data collection for the analysis about leadership and stakeholder management was conducted through retrospective interviews, where interviewers may have omitted or filtered details. Nevertheless, data triangulation and data collection at different moments allowed a reduction of this risk. An additional limitation is that this study was focused on the role of leaders in involving library staff and internal stakeholders within the university. External stakeholders such as suppliers, other libraries and government institutions were not analysed. Finally, the last limitation is that the change process analysis was performed until December 2012; however, the change process is still in progress.

6. CONCLUSIONS

Transforming a traditional library into a Library 2.0 has been investigated widely, often with a strong emphasis on the technological aspects. This paper illustrates the importance of the "human factor" in technological innovations. It was found that transformational leadership and stakeholder management play a critical role. To ensure the success of change, leaders have to align the vision and the implementation strategies of change among stakeholders at multiple levels. In this particular case, although library managers recognized the importance of different stakeholders, findings show that there is no specific strategy to manage them. In fact, stakeholder management actions in this case study were reactive and unsystematic. Library managers should act as transformational leaders creating sustainable and trustful relationships not only with the library staff but also with other stakeholders to reach this goal.

ACKNOWLEDGMENTS

We wish to thank the staff of the Regional Documentation Centre at University of Cuenca for their cooperation throughout this study. Special thanks to the Flemish Interuniversity Council (VLIR), the University of Cuenca, and the National Secretariat of Higher Education, Science, Technology and Innovation of Ecuador (SENESCYT) for supporting this research project.

REFERENCES

1. Albritton, R.L. (1998). A new paradigm of leader effectiveness for academic libraries: An empirical study of the Bass (1985) model of transformational leadership. In T. Mech & G. B. McCabe (Eds.), Leadership and academic librarians (pp. 66–82). West-port: Greenwood Press

2. Antonakis, J., Avolio, B.J., & Sivasubramaniam, N. (2003). Context and leadership: an examination of the nine-factor full-range leadership theory using the Multifactor Leadership Questionnaire. The Leadership Quarterly, 14(3), 261–295.

3. Armenakis, A.A., & Bedeian, A.G. (1999). Organizational change: A review of theory and research in the 1990s. Journal of Management, 25(3), 293–315. Retrieved July 16, 2014, from http://www.scribd.com/doc/189456956/Organizational-Change-A-Review-of-Theory-and-Research-in-the-1990s

4. Avolio, B.J., & Bass, B.M. (1991). The full range of leadership development: Basic and advanced manuals. Binghamton, NY: Bass, Avolio & Associates

5. Bass, B.M. (1990). From transactional to transformational leadership: Learning to share the vision. Organizational Dynamics, 18(3), 19–31. Retrieved July 16, 2014, from

http://strandtheory.org/images/From_transactional_to_transformational_-_Bass.pdf

6. Bass, B.M., & Riggio, R.E. (2006). Transformational leadership (2nd ed.). Mahwah, NJ: Lawrence Erlbaum

7. Bryson, J. (2004). What to do when stakeholders matter. Stakeholder identification and analysis techniques. Public Management Review, 6(1), 21–53. Retrieved July 16, 2014, from https://www.hhh.umn.edu/people/jmbryson/pdf/stakeholder_identification_analysis_techniques.pdf

8. Budd, J. (2005). The changing academic library: Operations, cultures, environments (Vol. 56). Chicago: Association of College and Research Libraries

9. Castiglione, J. (2006). Organizational learning and transformational leadership in the library environment. Library Management, 27(4/5), 289–299

10. Crosby, P.B. (1996). The absolutes of leadership. San Francisco: Jossey-Bass Publishers

11. Cummings, T.G., & Worley, C.G. (2005). Organization development and change (8th ed.). Mason, OH: Thomson/South-Western

12. De Bruyn, H., & Kruger, S. (2001). Strategic management: A southern African perspective (3 ed.). Johannesburg: Entrepro Publishers

13. Freeman, E., & McVea, J. (2001). A stakeholder approach to strategic management. Darden Business School Working Paper No. 01-02. Retrieved July 16, 2014, from http://papers.ssrn.com/sol3/papers.cfm?abstract_id=263511

14. Freeman, R.E. (1984). Strategic management: a stakeholder approach. Boston, MA: Pitman

15. Govender, P., Moodley, S., & Brijball Parumasur, S. (2005). The management of change at selected higher education institutions: An exploration of the critical ingredients for effective change management. Journal of Industrial Psychology, 31(1), 78–88. Retrieved July 16, 2014, from http://www.sajip.co.za/index.php/sajip/article/viewFile/186/183

16. Graetz, F., & Smith, A.C. (2009). Duality theory and organizing forms in change management. Journal of Change Management, 9(1), 9–25.

17. Holm-Nielsen, L.B., Thorn, K., Brunner, J.J., & Balán, J. (2005). Regional and international challenges to higher education in Latin America. In H. De Wit, C. Jaramillo, J. Gacel-Ávila, & J. Knight (Eds.), Higher Education in Latin America. The international Dimension (pp. 39–69). Washington, DC: The World Bank. Retrieved July 16, 2014, from http://siteresources.worldbank.org/EXTLACREGTOPEDUCATION/Resources/Higher_Ed_in_LAC_Intnal_Dimension.pdf

18. Horn, A. (2008). Strategic competence: to soar above. Library Management, 29(1/2), 5–17.

19. Huang, Y.-C., & Liao, L.-C. (2011). A college leader's transformational leadership. Journal of Management Research, 3(2), 1–22. Retrieved July 16, 2014, from http://www.macrothink.org/journal/index.php/jmr/article/view/646/483

20. Jantz, R.C. (2012). Innovation in academic libraries: An analysis of university librarians' perspectives. Library & Information Science Research, 34(1), 3–12. Jorgensen, D.L. (1989).

Participant observation: A methodology for human studies (Vol. 15). Newbury Park, CA: Sage

21. Kotter, J.P. (1995). Leading change - Why transformation efforts fail. Harvard Business Review, 73(2), 59–67. Reprinted in Best of HBR (2006), Retrieved July 16, 2014, from https://secureweb.mcgill.ca/hr/sites/mcgill.ca.hr/files/leading_change_john_kotter.pdf

22. Kotter, J.P. (1996). Leading change. Boston, MA: Harvard Business School Press

23. Kotter, J.P., & Cohen, D.S. (2002). The heart of change: real-life stories of how people change their organizations. Boston, MA: Harvard Business School Press

24. Kouzes, J.M., & Posner, B.Z. (2012). The leadership challenge: How to make extraordinary things happen (5th ed.). San Francisco, CA: Jossey-Bass

25. Laplume, A.O., Sonpar, K., & Litz, R.A. (2008). Stakeholder theory: Reviewing a theory that moves us. Journal of Management, 34(6), 1152–1189

26. López, J.J., & Vargas, J.G. (2012). Ambigüedad organizacional en la planeación estratégica. Revista de Administração FACES Journal, 11(2), 44–67. Retrieved July 16, 2014, from http://www.fumec.br/revistas/facesp/article/view/1215/837

27. López Segrera, F. (2008). Tendencias de la educación superior en el mundo y en América Latina y el Caribe. Avaliação: Revista da Avaliação da Educação Superior (Campinas), 13, 267–291

28. Marshall, S. (2008). Leading change at the middle: Stories of higher education middle leaders 'success'. (Unpublished thesis submitted in partial fulfillment of the degree of Master of Educational Leadership and Management), Unitec Institute of

Technology, New Zealand. Retrieved October 1, 2014, from http://hdl.handle.net/10652/1304

29. McGuigan, G.S. (2012). Addressing change in academic libraries: a review of classical organizational theory and implications for academic libraries. Library Philosophy and Practice 2012, paper 755. Retrieved July 16, 2014, from http://digitalcommons.unl.edu/cgi/viewcontent.cgi?article=1829&context=libphilprac

30. Miles, M.B., & Huberman, A.M. (1994). Qualitative data analysis: an expanded sourcebook. Los Angeles: Sage Publications

31. Mitchell, R.K., Agle, B.R., & Wood, D.J. (1997). Toward a theory of stakeholder identification and salience: Defining the principle of who and what really counts. Academy of Management Journal, 22(4), 853–886. Retrieved July 16, 2014, from http://courses.washington.edu/ilis580/readings/Mitchell_et_al_1997.pdf

32. Munatsi, R. (2010). Implementation of library 2.0 services in African academic and research libraries: need for fundamental rethink. In P. Jain, & R. Nfila (Eds.), Proceedings of the 19th Standing Conference of Eastern, Central, Southern Africa Library and Information Associations (SCESSAL), held at University of Botswana, Gaborone, 6–9 December 2010. Gaborone: Botswana Library Association

33. Riggs, D.E. (2001). The crisis and opportunities in library leadership. Journal of Library Administration, 32(3–4), 5–17.

34. Savage, G.T., Nix, T.W., Whitehead, C.J., & Blair, J.D. (1991). Strategies for assessing and managing organizational stakeholders. The Executive, 5(2), 61–75

35. Solís, P., & López, B. (2000). El concepto de las anarquías organizadas en el análisis organizacional. Contaduría y

Administración, 197, 35–45. Retrieved July 16, 2014, from http://www.ejournal.unam.mx/rca/197/RCA19704.pdf

36. Stavridis, S., & Tsimpoglou, F. (2012). EFQM in academic libraries: The application of a quality management model at the University of Cyprus library. Liber Quarterly, 22(1), 64–77. Retrieved July 16, 2014, from http://liber.library.uu.nl/index.php/lq/article/view/8041/8413

37. Stephens, D., & Russell, K. (2004). Organizational development, leadership, change, and the future of libraries. Library Trends, 53(1), 238–257. Retrieved July 16, 2014, from http://cdigital.uv.mx/bitstream/123456789/5984/2/Liderazgo%202.pdf

38. Stringer, R. (2002). Leadership and organizational climate: the cloud chamber effect. Upper Saddle River, NJ: Prentice Hall

39. Sucozhañay, D., Sigüenza-Guzman, L., Zhimnay, C., Cattrysse, D., Euwema, M., & De Witte, K. (2011). Change management strategies for successful implementation of a library 2.0. A case study in Ecuador. In L. Gómez Chova, I. Candel Torres, & A. López Martínez (Eds.), Proceedings of the 5th International Technology, Education and Development Conference (INTED 2011) (pp. 1051–1060). Valencia, Spain: International Association of Technology, Education and Development, IATED

40. Sukram, P., & Hoskins, R. (2012). Leading the way : the managerial skills of library managers in South African university libraries. Innovation, 45, 94–119

41. Svendsen, A. (1998). The stakeholder strategy: profiting from collaborative business relationships. San Francisco, CA: Berrett-Koehler Publishers

42. Swanborn, P. (2010). Case study research: What, why and how? Los Angeles: Sage

43. Tam, L., & Robertson, A. (2002). Managing change: libraries and information services in the digital age. Library Management, 23(8/9), 369–377. doi: 10.1108/01435120210439843. Retrieved July 16, 2014, from http://hub.hku.hk/bitstream/10722/48727/1/84942.pdf?accept=1

44. Torres, C.A., & Schugurensky, D. (2002). The political economy of higher education in the era of neoliberal globalization: Latin America in comparative perspective. Higher Education, 43(4), 429–455.

45. Turner, D.-M. (2007). Clinician readiness for transition to a fully integrated electronic health care delivery system. (PhD. dissertation), Walden University, Minnesota. Available from Dissertations & Theses: A&I. (Publication No. AAT 3247595)

46. vanDuinkerken, W., & Mosley, P.A. (2011). The challenge of library management: Leading with emotional engagement. Chicago: American Library Association

47. Verschueren, M. (2014). Leadership styles of head nurses and quality of care. (PhD), University of Leuven, Leuven

48. Williamson, V. (2008). Relationships and engagement: The challenges and opportunities for effective leadership and change management in a Canadian research library. Library Management, 29(1/2), 29–40. doi: 10.1108/01435120810844621

49. Yin, R. (2009). Case study research. Design and methods (4th ed. Vol. 5). California: Sage

50. Yukl, G. (1999). An evaluation of conceptual weaknesses in transformational and charismatic leadership theories. The Leadership Quarterly, 10(2), 285–305.

51. Yukl, G.A. (2013). Leadership in organizations (8th ed.). England: Pearson Education, Limited

APPENDIX 1: SEMI-STRUCTURAL INTERVIEW GUIDE TO STAKEHOLDERS

1. Description of the change process

1.1 In general, what did the change process consist of?

1.2 What was the problem or goal that was intended to solve or achieve?

1.3 Could you describe how the change process (planning, implementation, assessment, etc.) was held?

1.4 What were or are your interests / aspirations / concerns on this change? Did you agree with the change?

1.5 How did you participate in this process?

2. Analysis of stakeholder management

2.1 How did you get involved in this process? Who contacted you? What was/is the role that you played/play in the process?

2.2 Why do you think you were involved in this process? At what point of the process you got involved?

2.3 What other stakeholders were involved? Could you enlist and analyse them in terms of their role in the process and their power to influence it?

2.4 Do you know how these actors were involved? Why were they involved?

2.4.1 Do you know if there was a stakeholders' analysis to decide their involvement? What were the criteria used for such analysis?

2.5 How was the relationship between the actors? Was there any conflict?

3. Leadership

3.1 Who exercised leadership in this process of change? Why do you consider him or her as a leader?

3.2 What behaviour did the leader exhibit? What activities did the leader perform? How did the leader make his/her decisions?

3.3 How was your relationship with the leader? How was the leader's relationship with the other actors?

3.4 What type of communication was used between the leader and the other stakeholders?

3.5 From your perspective, how did leadership influence the change process and the results achieved?

4. Change process results

4.1 What were the results achieved? Why? What results were not achieved? Who were the key stakeholders to achieve these results? Why?

4.2 What do you think were the main problems / bottlenecks / resistance occurred in this process?

4.3 How did the organizational structure of the university influence the implementation of change?

4.4 What recommendations would you make to improve the management of change in universities?

Appendix 2: List of codes used for the analysis

Category	Codes	Sub-codes
Change Process (CP)	CP_ Change Context (CC)	CC_University context CC_Library context
	CP_Time line CP_Planning CP_Objectives CP_Implementation CP_Results CP_Evaluation	CP_First Period* CP_Second Period* CP_Third Period* CP_Fourth Period*
	CP_Responses to change (RC)	RC_Resistance RC_Willingness RC_Ambivalence
	CP_Setbacks*	
	CP_Hits*	
Leadership (L)	L_Leaders identified	L_Leader period 1-2* L_Leader period 3* L_Leader period 4* L_University's top administrators * L_Institutional Change Project*
	L_Leaders behaviours	
	L_Expected role of leaders*	
Transformational leadership (TL)	TL_Behaviours	TL_Intellectual stimulation TL_Idealized influence TL_Inspirational motivation TL_Individualized consideration
	TL_Effectiveness	
Stakeholder management (SM)	SM_Identification of stakeholder	SM: Identification strategies SM: Stakeholders identified* SM: Characteristics of stakeholders*
	SM_Analysis of stakeholders	SM_Criteria used SM_Techniques used
	SM: Formulation and implementation of strategies	SM_strategies planned SM_strategies implemented
	SM: Evaluation of stakeholder management	
	SM_Other practices to manage stakeholders*	
	SM_Library Council*	
Leadership (L) + Stakeholder Management (SM)	L+SM: Relationship of the leader with stakeholders*	TL+SM: Relationship of the leader with stakeholders*

* Codes that emerged from data.

Appendix 3: Example of summarizing matrix: Display of activities, objectives and outcomes of the first period.

Period	Main activities	Objectives	Outcomes
March 2007 – March 2009	Diagnosis of the library carried out by two international experts. Visits to libraries at national and international level. Development and implementation of the plan of change called "Project for technological improvement". Development of a first draft of new library regulation and a departmentalization proposal. Purchase and installation of new technological equipment.	1. Technological improvements.	a) Improved Internet access through the increase of the number of computers; b) Training room in the Central library was built; c) Improved library network (bandwidth); d) Implementation of a unique web site for the library; e) Installation of a closed-circuit security in the Central library; f) New servers and storage equipment were installed.
		2. Staff and users training.	Library staff and users attended several workshops.
		3. Organizational objectives.	The organizational outcomes were mainly focused on the elaboration of the library regulation and a departmentalization proposal. Departmentalization proposal was not operationalized.

2

CHAPTER

Changing Role of Academic Librarians in Open Access Environment

Intekhab Alam

Librarian A.K.K. New Law Academy Pune, Maharashtra (India).

ABSTRACT

Application of information and communication technology is supporting various ways of scholarly communication. The transition from print to electronic and paid resources to open access resources has a great impact on information society and resulted open access movement. This paper enumerates various declarations on open access and discusses the impact of open access on libraries particularly on the role of academic librarians. On the basis of existing literature, an attempt has been made in this paper to understand transforming role of academic librarians and suggest new responsibilities in open access environment.

Keywords: Open Access, Open Access Movement, Academic Librarians, Role of Librarians in Open Access, Scholarly Publishing Literacy.

INTRODUCTION

In the last two decades information technology has radically changed social communication pattern and information behavior.1 This transition has affected both scholarly communication and the forms of scholarly publications. Since the advent of scholarly journals in the mid of seventeenth century, scholars have been seeking better ways to disseminate their outputs2 . With the advancement of information and communication technology, the mode of access to scholarly information has changed from print to electronic. The concept of e-journals consortia emerged as a subscription model to journals during the time due to the high cost of access to e-journals. In parallel with this technological development the amount of research itself has increased exponentially, coupled with the desire for wider dissemination of research result. From this scenario open access has arisen3 as an alternative to subscriptions as a business model for scholarly journals4 and has attracted a great deal of attention in recent years. The Open Access Movement has focused on free access to research and scholarship and changed the approach to disseminate it.

Open Access and Open Access Movement

Peter Suber described that "Open access (OA) literature is digital, online, free of charge, and free of most copyright and licensing restrictions"5 . Open access can also be defines as providing free access for all Internet users to scholarly information that is published or to be published in research journals. Many academic libraries are trying to provide free and open access material in the form of scholarly journal publications, textbooks and open educational resources. While there are many free resources, the term "open access" refers specifically to content that is peer-reviewed and shareable according to author licenses, such as

Creative Commons, or it refers to contracts the author signs with a publisher to retain certain rights. These rights are listed on the SHERPA/RoMEO Web site (Securing a Hybrid Environment for Research Preservation/Rights Metadata for Open Archiving) and pertain to publisher policies on the self-archiving of journal articles online.6

Open Access Movement is more about growing consciousness among public towards having access to research results which are mostly supported by public money collected in the shape of taxes and other forms. The question arises, why should public pay for those things which already have come up with public money and making available research results to public is more of a matter of right, if it is supported by public money.7

The Open Access Movement was formally initiated by the "Budapest Open Access Initiative"8 on February 14, 2002 and further supported by "Bethesda Statement on Open Access Publishing"9 in June 2003 and "Berlin Declaration on Open Access to Knowledge in the Sciences and Humanities"10 in October 2003.

In Bethesda Statement, open access was redefined and again in Berlin Declaration the definition was refined and declared that open access contributions must satisfy two conditions:

1 The author(s) and right holder(s) of such contributions grant(s) to all users a free, irrevocable, worldwide, right of access to, and a license to copy, use, distribute, transmit and display the work publicly and to make and distribute derivative works, in any digital medium for any responsible purpose, subject to proper attribution of authorship (community standards, will continue to provide the mechanism for enforcement of proper attribution and responsible use of the published work, as they do now), as well as the right to make small numbers of printed copies for their personal use.

2. A complete version of the work and all supplemental materials, including a copy of the permission as stated above, in an appropriate standard electronic format is deposited (and thus published) in at least one online repository using suitable technical standards (such as the Open Archive definitions) that is supported and maintained by an academic institution, scholarly society, government agency, or other well established organization that seeks to enable open access, unrestricted distribution, inter operability, and long-term archiving.11

Other declarations on open access are:

1. Association of College & Research Libraries (ACRL) Principles and Strategies for the Reform of Scholarly Communication, August 28, 2003. (http://www.ala.org/acrl/publications/whitepapers/principlesstrategies)

2. UN World Summit on the Information Society (WSIS) Declaration of Principles and Plan Action, December 12, 2003. (http://www.itu.int/wsis/docs/promotional/brochure-dop-poa.pdf)

3. Organisation for Economic Co-operation and Development (OECD) Declaration on Access to Research Data from Public Funding, January 30, 2004. (http://www.oecd.org)

4. IFLA Statement on Open Access Scholarly Literature and Research Documentation, December 5, 2003. (http://archive.ifla.org/V/cdoc/open-access04.html)

5. On the basis of the Budapest, Bethesda, Berlin definitions it was analysed that:12

1. Open Access Literature is Freely Available
2. Open Access Literature is Online
3. Open Access Literature is Scholarly and Royalty Free
4. Open Access Literature can be used with Minimal Restrictions
5. Open Access Literature Metadata can be Harvested
6. Open Access Strategies: Self-Archiving and Open Access Journals

Open Access and Libraries

For libraries, open access offers a variety of new opportunities and challenges. In the long-term, open access may help shrinking materials budgets by mitigating costs for resource purchase and access. It may also reduce some of the headaches involved in negotiating electronic journals and database licenses.13 Many open access advocates are optimistic that it will enhance scholarly communication by making content broadly available and by providing relief to libraries from continuing large price increases of subscription journals.14 There is no doubt, open access has enhanced scholarly communication and library users are getting unrestricted and free access to scholarship. It is always not possible for a library to purchase or access to full range of research literature on payment basis while open access publishing model has added a lot to library collection indirectly. Libraries play an essential role in open access developments by their expertise in building infrastructure, in creating user-friendly services of high quality and in securing long-term access.15 Academic libraries have played an active role in the expansion of the OA movement by promoting it in a variety of ways:16

- » including records for OA journals in their public catalogues and e-journals lists
- » collaborating with their institutions to establish institutional repositories
- » participating in, and at times, leading institutional initiatives to encourage faculty to deposit their work in the institution's repository
- » digitizing historical collections and providing OA
- » becoming active OA journal publishers

Open Access and Role of Librarians

As new models of scholarly communication emerge, librarians are situated to play a key role in the development of these models for academic publishing and dissemination.17 In 2007, the working group of the National Knowledge Commission on 'Open Access and Open Educational Resources' recommended in their report to increase Open Access content from India, in order to increase the public awareness and utilization of OA material.18 Now it is the duty of librarians to do efforts for creating an environment for open access and make aware to the academic community to maximize the usage of open access resources available either in Institutional repositories or otherwise. Librarians have shown their support for open access by signing open access initiatives and petitions. They have educated faculty and administrators about evolving scholarly communication environment. Libraries have partnered with faculty and research managers to set up open access repositories and to help faculty and students to deposit their research outputs. Librarians have provided support in research data curation and sharing. They have helped scholarly publishers to publish open access journals and books, and they have worked with educators to produce open educational resources ensuring the quality of digital content, its reuse and sharing. Open access has thus changed the profile of academic and research libraries.19 The librarians' desire to acquire, organize, preserve and disseminate the information available freely is the need of the hour. It is a challenge to generate the interest of the librarians in providing greater access to the masses at lower cost. The librarians must be positive and they must always be eager to adopt new techniques to cope with the changes and ready to do adjustment with the rapidly evolving conditions.

RELATED LITERATURE

The first of its kind study conducted in the form of national survey in USA in 2006 to study the attitude of the academic librarians and certain actions about open access principles. It was resulted that academic librarians support the open access and their involvement is increasing in scholarly communication by establishing institutional repositories and other open access initiatives. The study could not reveal the personal actions of librarians but 30% librarians are with a view that academic libraries should create professional positions whose main duties concern open access issues and projects, and 49% had no opinion or only somewhat agree. Over half of the respondents reported never having spoken with non-library staff about open access.20 Another study on the same lines was conducted covering Nigerian Private Universities in 2009- 10 with the objectives to evaluate the perceptions and reactions of academic librarians. It shows that most of the librarians are involved with multiple assignments in the libraries and have a very positive attitude towards open access. They agreed that academic librarians should educate the academic community about the issues and facilities of open access. Regarding the role of academic libraries 73.8%

respondents are agreed that libraries should create links to open access resources in their websites while 92.9% agree that bibliographic records for open access journals should be included in their catalogues.21

To find out the extent librarians have appreciated the new culture of using open access and institutional repository in their scholarly publications, this study was conducted in South Eastern Nigeria. The questions were asked regarding advantages of open access, role of librarians in institutional repository, constraints to the use of open access in scholarly publishing and strategies to enhance open access for scholarly publications. The result reveals that majority of the librarians are aware of open access information and accepted all the advantages of open access but only some of them have published their article in open access journal. Respondents accept role (as listed) of librarians in institutional repository, however respondents of one university did not agree that librarian are familiar with vendor licensing and copyright laws. All the constraints to the use of open access scholarly publishing and strategies to enhance open access for scholarly publication as listed in questionnaire were accepted by the respondents. It was suggested to find out why many librarians do not publish in open access journals.22 Mercer conducted a survey to analyse and better understand whether librarians employ open access method to make their research available and to offer possible reasons for academic authorlibrarian behaviours and motivations, with an eye to increasing awareness of and activism in scholarly communication among this group. It was found that out of the 3873 articles indexed in LISA almost 49% of academic librarians' articles were available open access journals. This percentage was bigger than the one reported for articles published by other departments (40%). The author characterized librarians' output as promising.23

The efforts of librarians in promoting and advocacy for open access started in last decade. In recent years some academic libraries have started to raise awareness among the students community. Through an online survey distributed to three listservs – Electronic Resources in Libraries, Serials in Libraries Discussion Forum and American College and Research Libraries' Scholarly Communication, the study results provide some insights into the thinking of a subset of academic librarians regarding what is being done at their institutions. The questions were asked regarding the important aspects of OA to which users should be familiar, activities of the librarians to educate about OA, and opinion on the role of librarians to educate students about OA. 86% of the respondents indicated that their library provided a link to the DOAJ on their library website but 62% reported that their library was doing enough to make OA resources accessible on their websites. 59% believe that their libraries were not doing enough to promote OA to their students. The study also reveals various promotional and educational activities initiated by the librarians towards promoting OA.24

A leading open access publisher - InTech has published the findings of a survey appraising attitude and awareness of the library community towards the open access business model in scholarly publishing. The online survey was conducted in April 2012 with the objectives to explore current levels of familiarity with open access; current opinion of and attitude towards open access; current role of the librarian with regards to open access and whether and how open access is changing the role of the librarian. Majority of the respondents were from large academic institutions based in Europe and North America. Results shows that familiarity with open access is quite good (95%) among the librarian community 54% of respondents state their author and reader communities are not familiar with open access. 78% of respondents are in favour of open access and 95% think that open access is already bridging tangible benefits and will do so in the future. 70% of participants' institution already have an institutional repository and majority of the participants (97%) feels that librarians should provide information on open access. Majority of the librarians are agree that open access is changing the role of the librarian positively and librarians should involve themselves in the pertinent activities and acquire necessary skills.25

The SAGE hosted a roundtable in association of the British Library to discuss challenges and changes ahead for libraries in relation to the impacts of changes in scholarly communication on 26 April 2012. The aims of the roundtable were to provide an international perspective on the likely impact of an open access future on librarians, to identify support and skills required for librarians in such a future, and to further current discussion on support for the library community from their institutions, publishers, funders and other parties. The group of 14 librarians and other industry experts discussed a number of key questions like setting parameters for what the likely shift towards open access might be in different disciplines and different geographic regions, then considering what the impact of such a shift would mean for the academic library community. The Report of roundtable meeting suggested following actions:

- » academic libraries need to evolve to continue to meet the changing needs of their users
- » libraries will need to look for ways to work together; greater dialogue is needed between libraries about strategies for dealing with open access and best practices
- » clear communication of open access, its benefits, processes and implications is needed for researchers

- » a review should be taken exploring attitude towards the role of institutional collections as more content as available as OA
- » as with the publishing models, open access carries both benefits and cost. The group believes that good policy outcomes will only result if those involved in the marketplace are willing to acknowledge and evaluate both and calls upon those involved to maintain open dialogue on open access.26

Transforming Role of Librarians

With the advancement of information communication technology and its implementation in library operations and services, the role of librarians has been changed significantly. Open access supported by technology has become a public movement globally and this movement has provided new challenges as well as opportunities for the librarians. With knowledge of open access, understanding of copyright and licensing, expertise in bibliometrics and applying quality indicators for research quality evaluation, and access to a range of resources and tools, academic librarians are well situated to claim a proactive role in supporting scholarly publishing literacy in the following areas:27

1. Raising awareness of open access developments, for example the different roads to open access, operations of open access repositories, and hybrid and fully open access journals
2. Assisting researchers with accessing tools and resources to enable them to understand funders and publishers' policies related to open access, for example database such as SHERPA/RoMEO, SHERPA/JULIET
3. Supporting the management of authors' rights, for example information on copyright and licensing, including Creative Commons

4. Administering and promoting the use of institutional repositories, for example supporting the depositing of research outputs

5. Using bibliometrics tools and other journal quality indicators for quality and impact evaluation, due diligence checking on open access journals

The librarians are facing challenges and opportunities due to the impact of open access movement on libraries; they have to play important roles by their activities or by rendering new services to users to be relevant in the changing information landscape:

1. They should do efforts to educate academic community about the benefits of making their research openly available.

2. They should have positive attitude towards open access policies and open access movement.

3. Establishing institutional repositories, providing link to open access resources in library web pages and supporting in publication of open access journal, they can enhance the scholarly communication.

4. They should publish their own research work in open access journals/repositories to become participate in open access movement.

5. They have to do research on open access scholarly communication periodically to know the dependency of academic community on open access information resources and take policy decisions accordingly.

6. They need to improve their professional knowledge and skills to deal with various stakeholders in scholarly communication process.

7. They should prepare young staff by providing training to cope with the new challenges.

CONCLUSION

In the present Indian scenario, the roles of librarians are refreshed and even more specialized roles are defined. Apart from traditional roles, they are managing digital library projects, digital archiving, institutional repositories, publishing open access journal, providing metadata harvesting services, creating web portals etc. In addition to current practices by librarians, scholarly publishing literacy needs to be considered through the lenses of digital scholarship and information literacy to be fully understood.28 Though the roles of the librarians and support services provided by them differ from one library to another, it is the great responsibility of the library schools to produce librarians with sound theoretical knowledge and trained with new skills to respond to the challenges in the transforming profession.

REFERENCES

1. Trier, G.V. (2009), "Focus: Scholarly Publishing and Open Access", available at:
2. Palmer, K.L., Dill, E. and Christie, C. (2009), "Where There's a will There's a Way?: Survey of Academis Librarian Attitudes about Open Access", College & Research Libraries, Vol. 70 No. 4, pp. 315-335.
3. Jamieson, G. (2004), "Open Access: the Future of Academic Publishing?", Cardiology in the Young, Vol. 14 No. 5, pp. 471-472.

4. Lewis, D.W. (2012), "The Inevitability of Open Access", College & Research Libraries, Vol. 73 No.

5. , pp. 493-506. 5 Suber, P. (2012), Open Access, MIT Press, Cambridge.

6. Martin, R. A. (2010), "Finding Free and Open Access Resources: A Value-Added Services for Patrons", Journal of Interlibrary Loan, Document Delivery & Electronic Reserve, Vol. 20 No. 3, pp. 189-200.

7. Pandita, R. and Ramesha (2013), "Open Access: A Global Public Movement", Journal of Indian Library Association, Vol. 49 No. 3, pp. 35-39.

8. BOAI (2002), "Budapest Open Access Initiative", available at: http://www.opensocietyfoundations.org/openaccess/read (accessed 2 August 2014).

9. BSOAP (2003), "Bethesda Statement on Open Access Publishing", available at: http://legacy.earlham.edu/~peters/fos/bethesda.htm (accessed 2 August 2014).

10. BDOAKSH (2003), "Berlin Declaration on Open Access to Knowledge in the Sciences and Humanities", available at: http://www.zim.mpg.de/openaccess-berlin/berlin_declaration.pdf (accessed 2 August 2014).

11. Ibid.

12. Bailey, C.W., Jr. (2007), "Open Access and Libraries", Collection Management, Vol.32 No. 3/4, pp. 351-383.

13. Supra note 2.

14. Supra note 1.

15. IFLA (2012), "IFLA Statement on Open Access", Italian Journal of Library and Information Science, Vol. 3 No. 2, pp. 1-12, available at: http://leo.cilea.it/index.php/jlis/article/view/8639/7774 (accessed 2 August 2014).

16. Cryer, E. and Collins, M. (2011), "Incorporating Open Access into Libraries", Serials Review, Vol. 37 No. 2, pp. 103-107.

17. Richard, J., Koufogiannakis, D. and Ryan, P. (2009), "Librarians and Libraries Supporting Open Access Publishing", Canadian Journal of Higher Education, Vol. 39 No. 3, pp. 33-48.

18. Sawant, S. (2013), "Past and Present Scenario of Open Access Movement in India", The Journal of Academic Librarianship, Vol. 39 No. 1, pp. 108-109.

19. Supra note 15.

20. Supra note 2

21. Bolarinwa, O. and Utulu, S. C. A. (2011), "Open Access: Perceptions and Reactions of Academic Librarians in Nigerian Private Universities", African Journal of Library & Information Science, Vol. 21 No. 2, pp. 121-131.

22. Okoye, M. O. and Ejikeme, A. N. (2011), "Open Access, Institutional Repositories and Scholarly Publishing: The Role of Librarians in South Eastern Nigeria", available at http://www.webpages.uidaho.edu/~mbolin/okoye-ejikeme.htm (accessed 3 August 2014).

23. Mercer, H. (2011), "Almost Halfway There: An Analysis of the Open Access Behaviour of Academic Librarians", College & Research Libraries, Vol. 72 No. 5, pp. 443-453.

24. Keane, E. P. (2012), "Librarian Viewpoints on Teaching Open Access Publishing Principles to College Students", The Serials Librarian, Vol. 63, pp. 333-349.

25. INTECH (2012), "Assessing the role of Librarian in an Open Access World", available at:http://www.intechopen.com/js/ckeditor/kcfinder/upload/files/Role%20of%20the%20Librarian_Survey_Findings_Jun12.pdf (accessed 3 August 2014).

26. Harris, S. (2012), "Moving Towards an Open Access Future: The Role of Academic Librarries", available at: http://www.uk.sagepub.com/repository/binaries/pdf/LibraryOAReport.pdf (accessed 3 August 2014).

27. ZHAO, L. (2014), "Riding the Wave of Open Access: Providing Library Research Support for Scholarly Publishing Literacy", Australian Academic & Research Libraries, Vol. 45 No. 1, pp. 3-18.

28. Ibid.

3

CHAPTER

Improving Library Management by Using Cost Analysis Tools: A Case Study for Cataloguing Processes

Lorena Siguenza-Guzman , Alexandra Van den Abbeele, Dirk Cattrysse

Centre for Industrial Management Traffic & Infrastructure, Faculty of Engineering, KU Leuven (BELGIUM) Faculty of Engineering, University of Cuenca (ECUADOR), BE

Faculty of Business and Economics, KU Leuven (BELGIUM), BE

Centre for Industrial Management Traffic & Infrastructure, Faculty of Engineering, KU Leuven (BELGIUM), BE

ABSTRACT

TTDABC is a relatively new costing management technique, initially developed for manufacturing processes, which is gaining attention in libraries. This is because TDABC is a fast and simple method that only requires two parameters, an estimation of time required to perform an activity and the unit cost per time of supplying capacity. A few case studies have been documented with regard to TDABC in libraries; all of them being oriented to analyse specific library activities such as inter-library loan, acquisition and circulation processes. The primary focus of this paper is to describe TDABC implementation in one of the most important library processes, namely cataloguing. In particular, original and copy cataloguing are analysed through a case study to demonstrate the applicability and usefulness of TDABC to perform cost analysis of cataloguing processes.

Keywords: Costing analysis, Time-Driven Activity-Based Costing, Cataloguing Processes, Academic Library

1. INTRODUCTION

In the current economic situation, characterized by periodic shortages and limited budget resources, libraries are in search of methods to improve their process efficiency and provide high-quality services at lower costs (**ACRL Research Planning and Review Committee, 2010**; **Cottrell, 2012**). In order to improve their performance, library managers need to consider cost reduction and the inclusion of service costs in their decisions (**Hoozée, Vermeire, & Bruggeman, 2012**). Furthermore, they should strive to identify improvement opportunities and to eliminate costs related to non-value adding activities (**Ellis-Newman, Izan, & Robinson, 1996**). In order to do this, library managers must keep their activities, resources and costs under control, relying on valid information about activity costs, resource capacity and their performance (**Stouthuysen, Swiggers, Reheul, & Roodhooft, 2010**).

In service institutions such as libraries, several methodologies for costing analysis have been used for decades, whereas the Time-Driven Activity-Based Costing (TDABC) is one of the most recent approaches. TDABC is a cost management technique developed by Kaplan and Anderson to overcome the difficulties presented by previous costing systems (**Kaplan & Anderson, 2007**). By implementing TDABC in libraries, key benefits are expected, including the possibility of benchmarking different scenarios; identifying non-value added activities; and justifying decisions and choices for staff recruitment, training and new service development (**Siguenza-Guzman, Van den Abbeele, Vandewalle, Verhaaren, & Cattrysse, 2013a**).

Although some research has been carried out with respect to TDABC in libraries, these studies have been focused on specific library activities such as the inter-library loan (ILL), acquisition, and circulation (**Siguenza-Guzman, Van den Abbeele, Vandewalle, Verhaaren, & Cattrysse, 2013b**). More research is still required to determine whether TDABC can be useful in other library services. Therefore, the aim of the paper is to provide more detailed insight on implementing TDABC for library cataloguing processes. We focus on this unit because cataloguing is considered for many libraries, to be one of their most expensive processes, especially original cataloguing (**Manaf & Rahman, 2006**). The remainder of this paper is organized as follows. Firstly, a brief outline of the theoretical background of costing systems is provided (§ 2). Secondly, the different steps involved in implementing TDABC in original and copy cataloguing are explained (§ 3). Thirdly, we discuss the similarities and differences found in the original and copy cataloguing regarding time and cost per activity (§ 4). In addition, a number of recommendations where process improvements were unveiled are included in this section. Fourthly, the benefits encountered when implementing the TDABC model in an academic library in Belgium are described (§ 5). We end this paper with a brief conclusion (§ 6).

2. THEORETICAL BACKGROUND: COSTING SYSTEMS

2.1. Traditional Costing Systems

Several library cost analysis studies have been performed since the 1970s (**Roberts, 2003**). However, these studies were mainly treated as technical rather than organizational or managerial innovations (**Kont, 2011**). **Ellis-Newman et al. (1996)**report that the majority of prior studies on library costs were undertaken in the United States, utilizing cost allocation models more compatible with traditional costing methods. Traditionally, the total product cost consists of direct costs, such as the cost of materials and direct labour, and a percentage of overheads as indirect costs (**Siguenza-Guzman et al., 2013a**). The latter includes training, marketing, and infrastructure, among others. Traditional costing systems are adequate when indirect expenses are low and product variety is limited. However, in an environment with a broad range of products and enhanced services, such as a library, indirect costs have become substantially more complex than direct costs (**Siguenza-Guzman et al., 2013a**). This situation renders traditional methods inadequate; not only for estimating the effect of strategic decisions, but also for providing crucial information to library managers (**Ellis-Newman & Robinson, 1998**; **Kaplan & Cooper, 1998**).

2.2. Activity-Based Costing Systems

Activity-Based Costing (ABC) is an advanced costing calculation that seeks to remedy the limitations of traditional methods (**Kaplan & Cooper, 1998**). ABC, promoted by Cooper and Kaplan in the mid-1980s, first accumulates indirect costs for each activity and then assigns the activity costs to the services causing that activity (**Cooper & Kaplan, 1988;Ellis-Newman & Robinson, 1998**). ABC has proven to be a valuable tool for libraries through its implementation in several case studies. For instance, **Ellis-Newman et al. (1996)** examined the application of ABC in academic libraries of two Western Australian universities. This study illustrates how activity costing helps managers to differentiate key activities from others that do not add value. In an additional study, **Ellis-Newman and Robinson (1998)** discuss the benefits of ABC for library managers and the steps involved in implementing ABC in an academic library. The authors show how traditional costing systems are unable to explain the relationship between costs and the underlying activities. Furthermore, **Ellis-Newman (2003)** demonstrates the type of information that an ABC system provides to assist decision-making with a case study in the user services area of an Australian academic library. In turn, **Ching, Leung, Fidow and Huang (2008)**employ ABC to examine the Super e-Book Consortium in Taiwan and Hong Kong. The study finds cost drivers of consortia business operations, and identifies the key consortium activities and their relevant costs. Moreover, **Goddard and Ooi (1998)**examined the development of ABC through a case study applied to library services at the University of Southampton. The authors present ABC as an option to overcome some of the problems of overhead allocations. Despite the benefits of ABC, they also describe significant problems with its practical application such as the amount of resources and time required for its development and maintenance. Eventually, **Novak, Paulus and Clair (2011)** describe how a medium-sized university library implemented ABC and other decision-making strategies to make budgetary cuts and thereby, redirecting library services.

Although a relatively extensive stream of literature finds that ABC systems provide interesting advantages for decision making in libraries, ABC also has its limitations.**Kaplan and Anderson** (**2004**, **2007**) note that ABC is difficult and costly to implement and maintain, especially when the current accounting system does not support the collection of ABC information. Data collection is time consuming and costly because of the need to interview and survey the library staff to estimate the percentage of time spent on each activity. While it works well in small organizations and limited activities, it becomes problematic to scale up to larger organizations (**Hoozée et al., 2012**). Managers also question the accuracy of the system since cost assignments are based on individuals' subjective information on how they spend their time. Furthermore, staff resistance could arise, as employees might feel threatened by the suggestion that their work should be improved. As a consequence, ABC systems tend to be outdated, and in some cases are abandoned and substituted by less demanding approaches such as Time-Driven Activity-Based Costing (**Wegmann & Nozile, 2009**; **Yilmaz, 2008**).

2.3. Time-Driven Activity-Based Costing Systems

TDABC is a useful cost management technique developed by **Kaplan and Anderson in 2004** to overcome the difficulties presented by previous costing systems (**Kaplan & Anderson, 2004**). TDABC assigns resource costs directly to cost objects using a fast and simple framework that only requires the unit cost of supplying resource capacity, and an estimation of the time duration of an activity (**Kaplan & Anderson, 2007**). Unlike the percentages that employees subjectively estimate for an ABC model, the time duration in a TDABC model can be readily observed and validated (**Kaplan & Anderson, 2007**). For each activity, «costing equations» are calculated and computed by time equations, which are the sum of individual activity times (**Yilmaz, 2008**). Through the use of time equations, TDABC allows incorporation of variation in the time demands made by different types of transactions, and consequently the representation of all possible combinations of activities that a process performs (**Kaplan & Anderson, 2007**). Five main TDABC advantages, highlighted by **Siguenza-Guzman et al. (2013a)**, are its simplicity to build accurate models and improve the understanding of the different processes; the opportunity of modelling complex operations thanks to the use of multiple drivers; a good estimation of resource consumption and capacity utilization; its fast maintenance compared to ABC models; and the possibility of using TDABC in a predictive manner.

TDABC has been carried out in specific library activities such as inter-library loan (ILL), acquisition and circulation. For instance, the first case study by **Pernot, Roothooft and Van den Abbeele (2007)** uses TDABC to calculate ILL costs and describes TDABC as a useful technique to reduce ILL resource costs and to renegotiate ILL service prices based on more accurate costs. The authors conclude that TDABC is very suited to cope with increasing cost pressures, and its findings can contribute to improve library services at lower costs. A second case study presented by **Stouthuysen et al. (2010)** describes the use of

TDABC for a library acquisition process. The authors state that TDABC provides library managers with a better insight into cost drivers, visualizes the acquisition process efficiencies and capacity utilization, and leads to potential cost efficiencies. As an illustration, they consider that 50% of some costs could be saved if administrative assistants get involved in the acquisition process instead of being performed by the head of department, provided that they are capable of doing these tasks. In addition, the authors demonstrate that TDABC can be updated rapidly and inexpensively to changes. Due to this flexibility, they consider that TDABC can be applied to other processes such as cataloguing or digitalized activities, with significant benefits. The latest case study by**Siguenza-Guzman, Van den Abbeele, Vandewalle, Verhaaren and Cattrysse (2014)** uses TDABC to analyse lending and returning processes. The authors provide several important insights for a successful implementation of TDABC in libraries such as: 1) collecting the time duration of the activities through direct observation to improve the level of accuracy; 2) using graphical representations of activity flows to validate the collected information straightforwardly; and 3) showing that clarifying the measurement purpose is crucial in improving the level of acceptance, and achieving the desired commitment with staff. They conclude that the TDABC implementation is worthwhile since it leads to a more accurate cost and process analysis for supporting decision-making.

3. TDABC IN CATALOGUING PROCESSES

The data used for this research was collected at the Arenberg Campus Library1(Campusbibliotheek Arenberg, hence the abbreviation CBA) of KU Leuven in Belgium. CBA offers information sources on subjects of the exact sciences, engineering, architecture, kinesiology, and rehabilitation sciences (**Campus Bibliotheek Arenberg, 2013**). Its services are handled by approximately 20.5 full-time equivalent employees (FTE). In this case study, we focus on describing the application of TDABC in two types of cataloguing activities: original and copy cataloguing. The former refers to creating a new bibliographic record from scratch, while the latter to adapting a pre-existing record to the characteristics of the item in hand (**Reitz, 2004**).

For this case study, qualitative interview data with quantitative data analysis were combined following the six steps presented by **Everaert, Bruggeman, Sarens, Anderson and Levant (2008)** to calculate the cost of activities through the TDABC model. These steps illustrated in Table 1 are described in detail by **Siguenza-Guzman et al. (2014)**.

Table 1 Time-Driven Activity-Based Costing steps (**Everaert et al., 2008**).

Time-Driven Activity-Based Costing	
Step 1	Identification of resource groups
Step 2	Estimation of the total cost of each resource group
Step 3	Estimation of the practical capacity of each resource group
Step 4	Calculation of the unit cost of each resource group
Step 5	Estimation of the standard time duration of each activity
Step 6	Multiplying the unit cost of each resource group by the time duration per activity

To identify resource groups involved in cataloguing activities as required in Step 1, multiple interviews were conducted. Initial interviews started through brief discussions with the library manager, and then moved to a more detailed level with the library staff. For each activity, a final interview was performed in order to validate specific details about the different sub-activities. During the interviews, key activities involved in each process were described in detail by employees in charge. The information was used to build flow charts of activity sequences. As **Siguenza-Guzman et al. (2014)** indicate, flowcharts allow a good overview of the different activities performed in a process, to identify additional expenditures such as computer maintenance and software licences, and afterwards to validate the activities in an optimal and simple manner. Figure 1 shows the activity flow of original and copy cataloguing respectively.

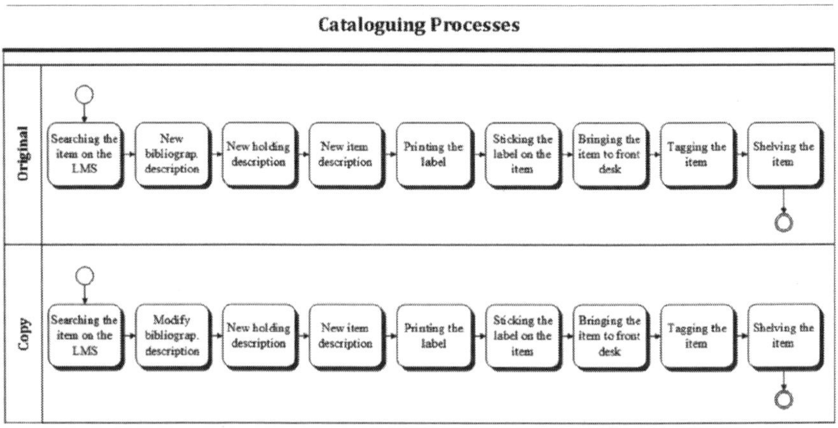

Note. – *LMS = Library Management System.*

Figure 1. Original cataloguing process.

The total cost of each resource group required in Step 2 was provided by the accountant and library manager via the Library Management System (LMS). The costs were classified into direct and indirect costs. Direct costs included salaries of staff and student library employees — SLE (i.e. students hired to perform secondary activities), equipment, and technology. Conversely, examples of indirect costs included stationery, electricity, support, telephone, training, and other items used to perform an activity (**Vazakidis & Karagiannis, 2009**).

The salary of cataloguers were calculated based on the average salary earned by employees responsible for cataloguing. According to the Chief Librarian, the total number of personnel assigned to cataloguing represents 1 full-time equivalent (FTE). The 1 FTE consisted of six people, each dedicating different amounts of their time to cataloguing processes. This corresponded to about €59,000 on a yearly basis. LMS costs covered the annual integrated library system and supporting software license fees attributed for the CBA. The Library Management System includes functionalities for acquisition, cataloguing, circulation and reporting. This integrated library solution was acquired for the entire library which consisted of the Central Library, three Campus Libraries and several Faculty Libraries for the Humanities and Social Sciences group. The LMS costs amounted to €17,000 on a yearly base. The annual computer maintenance costs for specific tasks such as reparation, maintenance, cleaning and depreciation of a PC in the cataloguing processes equaled €5,000. RFID maintenance costs refer to the costs associated with the maintenance, repair and inspection of the RFID system and yearly costs corresponded to about €17,000.

In the case of indirect costs, the library accountant estimated that about €195,000 was annually spent on general overhead (GO) costs. GO costs included management, secretary, accounting, training, staff meetings and stationery material. Other indirect values such as electricity, telephone, heating and transportation were not accounted as part of general overhead costs since they were paid by the University and not charged to the library (**Ellis-Newman & Robinson, 1998**). In order to calculate the overhead costs attributed to cataloguing activities, general overhead cost was divided by the total number of FTE working at the entire library. This resulted in a yearly overhead of approximately €9,500 per FTE. An overview of the total cost of each resource group can be seen in Table 2.

Table 2 Total cost of each resource group.

Total cost per resource group	
Resource Group	Cost (€) per year
Cataloguers	€59,000
Library Management System	€17,000
Computer Maintenance	€5,000
RFID Maintenance	€17,000
General Overhead	€9,500

Then, the practical capacity estimation of each resource group was calculated in Step 3. According to **Kaplan and Anderson (2007**, pp. 52–53), practical time capacity can be estimated in two different ways: 1) assuming an 80% of theoretical time capacity for people due to breaks, arrival and departure, training, meetings, and chitchats; and 85% for machines due to maintenance, repair, and scheduling fluctuations. 2) Calculating the real values according to the library situation, for example, available working hours, excluding holidays, meetings and training hours. In order to simplify the study, the first option was selected. For staff capacity, 38 hours per week were accounted as theoretical time capacity. It means 30.4 hours per week for practical capacity (staff practical capacity = 38 hours * 80%). Assuming fifty-two weeks per year, the practical capacity of a cataloguer is equal to 94,848 minutes per year $\left(30.4\frac{\text{hours}}{\text{week}}\times 52\frac{\text{weeks}}{\text{year}}\times 60\frac{\text{minutes}}{\text{hour}}\right)$. In the case of machines, the theoretical time capacity was set equal to the time in which cataloguers were available that is again 38 hours per week. Thus, the practical capacity for machines is 32.3 hours per week (machines practical capacity = 38 hours * 85%), and 100,776 minutes per year $\left(32.3\frac{\text{hours}}{\text{week}}\times 52\frac{\text{weeks}}{\text{year}}\times 60\frac{\text{minutes}}{\text{hour}}\right)$.

Once the practical capacity was obtained, the cost per unit time was calculated in Step 4 by dividing the total cost of a resource (Step 2) by the practical time capacity (Step 3). An overview of the resulting costs involved in this analysis is shown in Table 3.

Table 3: Costs involved in the analysis.

Total cost per minute for the different resource groups	
Resource Group	**Cost per minute (€/min)**
Cataloguers	0.62
Library Management System	0.17
Computer Maintenance	0.05
RFID Maintenance	0.17
General Overhead	0.10

$$\text{Cost per unit time} = \frac{\text{time cost of the resource}}{\text{practical capacity}}$$

For the fifth step, **Kaplan and Anderson (2007**, p. 30) recommend that the time required to perform an activity should be estimated based on standard times rather than actual times, since these times might reflect random variation, individual employee variation and nonrecurring factors. Moreover, the authors argue that precision is not critical and that a rough accuracy is sufficient because gross inaccuracies will be revealed either in unexpected surpluses or shortages of committed resources. This level of accuracy can be obtained by multiple methods such as direct observation, interviews, process maps or leveraging time estimates from elsewhere in the institution (**Kaplan & Anderson, 2007**, p. 26). In our case study, the standard time to perform an activity was gathered in the academic period 2010–2011 through direct observation as recommended by **Siguenza-Guzman et al. (2014)**. Observations were made multiple times using a stopwatch during several days at different hours in order to avoid possible biases (**Siguenza-Guzman et al., 2014**). Finally, based on their average values, an additional interview was performed in order to validate the data collection. Next, time equations were constructed for each activity. Time equations are the sum of individual activity times, which are represented with the following expression (**Kaplan & Anderson, 2007**):

$$\text{Time required to perform an activity} = (\beta_0 + \beta_1 X_1 + \beta_2 X_2 + \beta_3 X_3 + \beta_4 X_4 + \beta_5 X_5 + \ldots + \beta_i X_i)$$

With:

β_0= The standard time to perform the basic activity (e.g. 2 minutes)

β_i= The estimated time for the incremental activity *i* (e.g. time required for a librarian to enter an item in the cataloguing system = 0.5 minutes)

$\mathbf{X}_i$= The quantity of incremental activity *i* (e.g. items per batch = 1, 2...)

3.1. Original Cataloguing

The process, as shown in Figure 1, starts by searching the item in hand on the LMS in order to verify whether a similar record is already present in the database. This searching takes on average of 57s. If the item and record do not appear to match, the cataloguer creates a new record which includes the bibliographic description, requiring 306s (or 5min and 6s). A bibliographic description is the standardized description of an item including: title, edition, material specific details, details of publication, standard number, etc. (**Reitz, 2004**). The cataloguer then creates a new holding description, which is usually the information concerning the location of an item, taking 32s; and finally a new item description in 15s. Item description indicates item type, volume number, barcode, and loan rules. Once the new record is processed and stored in the database, the cataloguer prints the corresponding label (30s), and sticks on the item in 28s. Afterwards, the cataloguer brings the item to the front desk (30s) in order to tag the item (40s). An individual tag costs €0.30 including VAT. Eventually, the item is placed on the corresponding shelf or stack, requiring 168s (or 2min and 48s).

3.2. Copy Cataloguing

In contrast to the above process as shown in Figure 2, the cataloguer does find a record that appears to match with the item in hand. The cataloguer requires 74s (or 1min and 14s) to validate and modify the bibliographic description. Next, the cataloguer creates a new holding and item description, taking 32s and 15s respectively. Labelling and shelving are the same as the original cataloguing process.

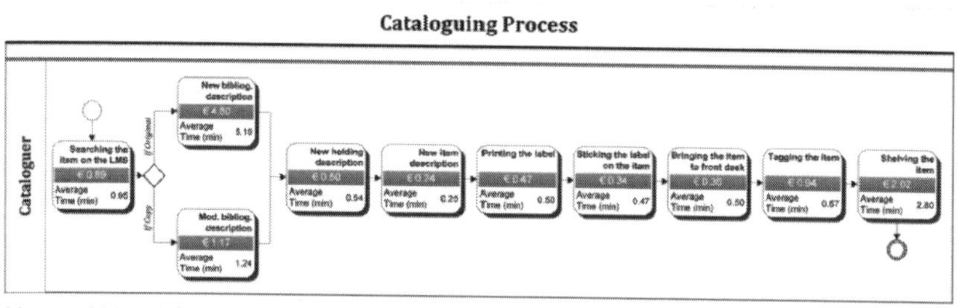

Note. – *LMS = Library Management System.*

Figure 2. Cataloguing process.

As the only difference between original and copy cataloguing is new versus modifying bibliographic description, we can create only one time equation by adding dummy variables in the equation. A dummy is a variable that takes the value 1 or 0 if a certain condition is true or false respectively. The resulting equation is as follows:

$$\begin{aligned}\boldsymbol{Cataloguing} &= \text{Searching} + \text{New}_{\text{Bibdescription}}\{\textit{if original\_cataloguing}\} \\ &+\text{Mod}_{\text{Bibdescription}}\{\textit{if copy\_cataloguing}\} + \text{New}_{\text{holding}} + \text{New}_{\text{item}} + \text{Print}_{\text{label}} \\ &+\text{Stick}_{\text{label}} + \text{Bring}_{\text{item}} + \text{Tag}_{\text{item}} + \text{Shelve}_{\text{item}}\end{aligned}$$

$$\begin{aligned}\boldsymbol{Cataloguing} &= 0.95 + 5.10\ \{\textit{if original\_cataloguing}\} + 1.24\{\textit{if copy\_cataloguing}\} \\ &+0.54 + 0.25 + 0.50 + 0.47 + 0.50 + 0.67 + 2.80\end{aligned}$$

Once the estimated time per activity and the unit cost of each resource group is calculated, costs are assigned to cost objects by multiplying the unit cost per time of resources by the estimated time required to perform the activity. It is represented by the following mathematical expression:

*Cost of an individual activity = time required to perform an activity * resources cost*

Figure 2 shows the resulting activity flow of the cataloguing process including average times and costs per sub-activity. Original and copy cataloguing are integrated in the same graphical representation by using decision diamonds. Finally, the total cost of the process is calculated by summing up all activity costs and can be represented by:

$$\text{Total Cost of a Process} = \sum \text{costs of individual activities}$$

The total costs and time incurred in the original and copy cataloguing processes are presented in Table 4. The resulting table is divided vertically into six columns and horizontally by standard and optional activities to separate the activities influenced by dummy variables. The first column lists the activities identified in the cataloguing process; the second column shows the average time per activity. The third column specifies the accumulated costs per minute of resources involved in each activity; the fourth column calculates the resulting cost incurred in the activity. The fifth column describes the condition under which two options are available to be selected: original and copy cataloguing. Eventually, the sixth column indicates the resource groups involved per activity.

Table 4 Total cost of the cataloguing process.

	Cataloguing Process					
	Activity	**Average Time (min)**	**Cost per min (€/min)**	**Cost (€)**	**Condition**	**Resources**
Standard Activities	Searching the item on the LMS	0.95	0.94	0.89		Cataloguer + LMS + CM + GO
	New holding description	0.54	0.94	0.50		Cataloguer + LMS + CM + GO
	New item description	0.25	0.94	0.24		Cataloguer + LMS + CM + GO
	Printing the label	0.50	0.94	0.47		Cataloguer + LMS + CM + GO
	Sticking the label on the item	0.47	0.72	0.34		Cataloguer + GO
	Bringing the item to front desk	0.50	0.72	0.36		Cataloguer + GO
	Tagging the item	0.67	1.41	0.94		Cataloguer + LMS + CM + GO + RM + Tag
	Shelving the item	2.80	0.72	2.02		Cataloguer + GO
	Subtotal	**6.68**		**5.77**		
Optional Activities	New bibliographic description	5.10	0.94	4.80	*if original_cataloguing*	Cataloguer + LMS + CM + GO
	Mod. bibliog. description	1.24	0.94	1.17	*if copy_cataloguing*	Cataloguer + LMS + CM + GO
	Subtotal	---		---		
	Total Original Cataloguing:	**11.78**		**10.57**		
	Total Copy Cataloguing	**7.92**		**6.94**		

Note. – LMS = Library Management System. GO = General Overhead. CM = Computer Maintenance. RM = RFID Maintenance.

The subtotal of "standard activities" is the sum of costs included in both processes: original and copy cataloguing. The average time in minutes for standard activities was 6.68 with an activity cost of €5.77. To calculate the total cost of the original cataloguing process, standard activities were summed up with the optional activity, new bibliographic description. The resulting time in minutes was 11.78 with a cost of €10.57. On the other hand, to calculate the total cost of the copy cataloguing process, standard activities were summed up with the optional activity, modify bibliographic description. The results show that the average time for copy cataloguing was 7.92 minutes and its cost was €6.94 per title.

4. ORIGINAL AND COPY CATALOGUING

Because of budgetary constraints and technological changes, cataloguing units have significantly influenced the nature of cataloguing work (**Mitchell, Thompson, & Wu, 2010**). In order to become more efficient, cataloguers constantly search for new ways to increase bibliographic access without spending more money (**Morris & Wool, 1999**). That includes automation, outsourcing, lowered costs for traditional cataloguing, and an increasing variety of information resources to control. With regard to lowering costs, a number of studies on cataloguing costs have been reported in the literature. For instance, the Iowa State University (ISU) conducted a longitudinal time-cost study to investigate the impact of automation on cataloguing costs (**Morris, Hobert, Osmus, & Wool, 2000**). Results showed that in 1987–1990, the average cost of original cataloguing at ISU was $34.13 and that copy cataloguing was $8.18 (**Morris, 1992**). In 1997–1998, original cataloguing costs performed by faculty cataloguers increased to $75.43 and to $58.72 when some original records were contributed by library assistants. Copy cataloguing costs, in the same period, increased to $35.82 when performed by faculty cataloguers and to $8.87 when performed by library assistants. Although costs of original and copy cataloguing increased during the time, **Morris et al. (2000)** also highlighted the fact that the average cost of cataloguing processes (i.e. copy, full original, minimal original and recataloguing) declined from $20.83 to $16.25 between 1990/91 and 1997/98. Authors attributed this decrease to national collaborative efforts, technological development, and reengineering efforts that have improved costs effectiveness and quality process (**Morris & Wool, 1999**). The University of Oregon, following the previous research at ISU, conducted a benchmark analysis during autumn 1997 to determine time and costs for acquisitions, cataloguing, and processing functions (**Slight–Gibney, 1999**). Results showed that the average cost for copy cataloguing was $9.23 per title, and that original cataloguing was $24.92. A third case study by **Ellis-Newman and Robinson (1998)** in an Australian academic library, reported the cost of library services utilizing ABC models. Results showed that the average cost of copy cataloguing at Edith Cowan University was $12.48 and original cataloguing was $54.39. Authors highlighted the importance of implementing activity-based costing systems in libraries to assign more accurate costs to services, to categorize costs, and to develop a price schedule for fee-based services.

In Table 5, all cases cataloguing costs vary library to library, even if the change in the rate of exchange and inflation are included in the three literature case studies.**Charbonneau (2005)** points out that these numbers logically vary because they are based on locally produced data and operations, individual cataloguing expertise, the type of material catalogued, and the cataloguing tools and resources available. Differences in overhead costs can also explain some of the cost variations. For instance, cataloguing overhead costs at ISU represent approximately 45% of the full costs, while in our case study about 10%. Unfortunately, not all studies provide detailed information on the overhead costs. In this case study, these costs include management, secretary, accounting, training, staff meetings and stationery material; however, other university costs such as electricity, telephone, heating and transportation are not accounted for since they are not charged to the library. At ISU overhead costs are organized into overhead centres: paid leave, automation and support services (**Morris et al., 2000**). Some other studies consider overhead costs as approximate since they are estimates or based on incomplete data (**Slight–Gibney, 1999**). Additional factors that vary among libraries are the library structure and process flow, for instance, pre-cataloguing activities are included in the cataloguing costs of one library (e.g. ISU), by the acquisitions department at another library, or even delegated to library assistants or students. In our case study, this activity is not included in the cataloguing flow because it is performed by the person responsible for acquisitions. Finally, differences in the periods of analysis can also explain such cataloguing costs variations, especially due to evolving factors that need to be taken into account. These factors include the increasing automation of cataloguing activities, use of shared cataloguing and authority records, decreasing staffing for cataloguing, and the growing presence of new information formats (**Morris & Wool, 1999**).

Table 5 Cataloguing costs adjusted using inflation and exchange-rates. † Cataloguing costs when performed by a library assistant.

Cataloguing costs in the four cases studies

University	Period	Original Cataloguing	Copy Cataloguing	Inflation	Inflated Original Cataloguing	Inflated Copy Cataloguing	EUR Conversion	Original Cataloguing (€)	Copy Cataloguing (€)
Iowa State University (US)	1987–1990	USD $34.13	USD $8.18	68.66%	57.56	13.80	0.72	41.39	9.92
Iowa State University (US)*	1997–1998	USD $75.43	USD $35.82	37.69%	103.86	49.32	0.72	74.68	35.46
Iowa State University (US)†	1997–1998	USD $58.72	USD $ 8.87	37.69%	80.85	12.21	0.72	58.13	8.78
University of Oregon (US)	1997	USD $24.92	USD $ 9.23	39.91%	34.87	12.91	0.72	25.07	9.29
Edith Cowan University (AU)	1998	AUD $54.39	AUD $12.48	47.20%	80.06	18.37	0.74	59.41	13.63
KU Leuven (BE)	2010–2011	EUR €10.57	EUR €6.94	-	-	-	-	10.46	6.83

Note.– Inflation data from inflationdata.com (US, Dec 1987/1998 to Dec 2011), and rba.gov.au (AU, 1998 to 2011).
Exchange data from x-rates.com (USD and AUD to EUR in 2011).
* Cataloguing costs when performed by a faculty cataloguer.
† Cataloguing costs when performed by a library assistant.

Although cataloguing costs data are not necessarily comparable among libraries (**McCain & Shorten, 2002**); "best practices" can be adopted in other libraries to influence their own workflows and to streamline their processes. This case study contributes to the cataloguing cost study literature by providing an additional approach for calculating cataloguing costs based on a fast and simple method as is TDABC. In this case study, the copy cataloguing, as shown in Figure 3, is approximately 34% less costly and 33% less time-consuming than original cataloguing. Thanks to the TDABC's ability to disaggregate costs per activity, it is possible to clearly analyse which activities demand more time, and thus lead to higher costs. For instance, the unique difference between original and copy cataloguing is the creation or modification of a bibliographic description. Results shown in Figure 4 indicate that adapting a bibliographic description from a pre-existing record is approximately 75% less costly than creating a new full bibliographic record. Based on these findings, it is concluded that is worthwhile to recommend librarians to adapt a pre-existing record instead of creating a new bibliographic record from scratch, as this will lead to significant costs and time reductions. Copy cataloguing not only increases cataloguers efficiency by eliminating duplication of effort, but also by reducing typographical errors caused in local libraries (**Beall & Kafadar, 2004**).

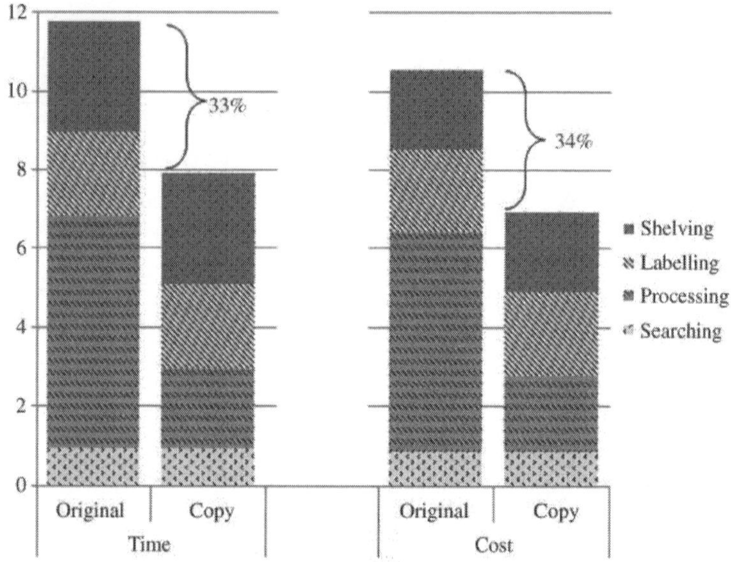

Figure 3. Comparison of original and copy cataloguing in terms of time and cost.

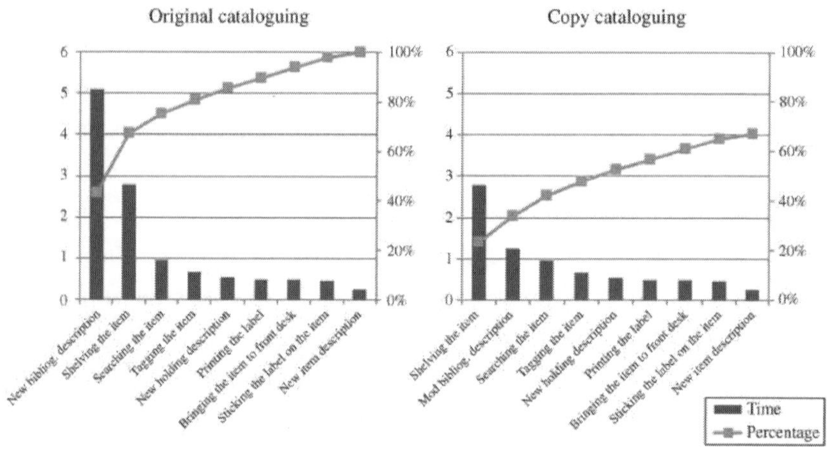

Figure 4. Pareto chart of disaggregating costs per activity for original and copy cataloguing.

Moreover, time equations show that in this particular case, labelling activities such as printing and sticking labels, bringing an item to the front desk, and tagging the item, consume approximately 20–30% of the cataloguing processes. Reviewing the consumed resources revealed that the cataloguer is typically in charge of labelling and shelving catalogued items. A "what-if analysis" can be performed to simulate the effect of delegating these activities to student library employees (SLE). If we take as reference the SLE costs calculated by **Siguenza-Guzman et al. (2014)** of 0.23 €/min, then the resulting costs would be €8.63 for original cataloguing, and €5.00 for copy cataloguing, as shown in Table 6. That is a cost reduction of about 18% for original cataloguing and 28% for copy cataloguing.

Table 6. Example of a what-if analysis applied to the original model of the cataloguing process.

	Cataloguing Process					
	Activity	Average Time (min)	Cost per min (€/min)	Cost (€)	Condition	Resources
Standard Activities	Searching the item on the LMS	0.95	0.94	0.89		Cataloguer + LMS + CM + GO
	New holding description	0.54	0.94	0.50		Cataloguer + LMS + CM + GO
	New item description	0.25	0.94	0.24		Cataloguer + LMS + CM + GO
	Printing the label	0.50	0.55	0.28		SLE + LMS + CM + GO
	Sticking the label on the item	0.47	0.33	0.15		SLE + GO
	Bringing the item to front desk	0.50	0.33	0.17		SLE + GO
	Tagging the item	0.67	1.02	0.68		SLE + LMS + CM + GO + RM + Tag
	Shelving the item	2.80	0.33	0.92		SLE + GO
	Subtotal	**6.68**		**3.83**		
Optional Activities	New bibliographic description	5.10	0.94	4.80	*if original_cataloguing*	Cataloguer + LMS + CM + GO
	Mod. bibliog. description	1.24	0.94	1.17	*if copy_cataloguing*	Cataloguer + LMS + CM + GO
	Subtotal	---		---		
	Total Original Cataloguing:	**11.78**		**8.63**		
	Total Copy Cataloguing	**7.92**		**5.00**		

Note. – LMS = Library Management System. SLE = Student Library Employees. GO = General Overhead. CM = Computer Maintenance. RM = RFID Maintenance.

An additional improvement to the process can be to incorporate batch cataloguing practices and mass retrieval rather than cataloguing individual items or records (**Mitchell et al., 2010**). The process flow analysis enables library managers to group and improve certain activities such as searching, processing (bibliographic, holding and item descriptions), labelling and shelving. These batch activities, such as processing a bunch of items or records at one time, can be performed based on the staff availability at certain moments of the day or week. An example of these sorts of improvements on the cataloguing process is shown in Table 7. In this case, searching, labelling, and shelving are delegated to SLEs and calculated in a batch of 10 items. The duration time of searching items needs to be augmented since SLEs have no expertise in the use of LMS whereas a cataloguer does have. The activity of bringing the item to the front desk is eliminated, suggesting the purchase of an extra rewriting RFID machine be located in the cataloguing computer. Finally, shelving activities were recalculated based on the batching times used by **Siguenza-Guzman et al. (2014)** to model a returning process flow. That includes, an SLE sorting the items in the cluster (i.e. book collection divisions); and then reshelving the items. The resulting time and cost after these improvements can be seen in Table 7.

Table 7. Example of a what-if analysis to improve applied to the original model of the cataloguing process

	Improvements on the Cataloguing Process					
	Activity	Average Time (min)	Cost per min (€/min)	Cost (€)	Condition	Resources
Searching	Searching the item on the LMS	1.21	0.55	0.67		SLE + LMS + CM + GO
Processing	Mod. bibliog. description	1.24	0.94	1.17	*if original_cataloguing*	Cataloguer + LMS + CM + GO
	New bibliographic description	5.10	0.94	4.80	*if copy_cataloguing*	Cataloguer + LMS + CM + GO
	New holding description	0.54	0.94	0.50		Cataloguer + LMS + CM + GO
	New item description	0.25	0.94	0.24		Cataloguer + LMS + CM + GO
Labelling	Printing the label	0.50	0.55	0.28		SLE + LMS + CM + GO
	Sticking the label on the item	0.47	0.33	0.15		SLE + GO
	Tagging the item	0.67	1.02	0.68		SLE + LMS + CM + GO + RM + Tag
Shelving*	Classifying the item	0.17	0.33	0.06		SLE + GO
	Shelving the item	0.35	0.33	0.12		SLE + GO
	Total Original Cataloguing:	9.26		7.49		
	Total Copy Cataloguing	5.40		3.85		

Note. – LMS = Library Management System. SLE = Student Library Employees. GO = General Overhead. CM = Computer Maintenance. RM = RFID Maintenance.
* Shelving time taken from Siguenza-Guzman et al. (2014).

Results of the what-if analysis indicate that by doing these changes, copy cataloguing is approximately 49% less costly and 42% less time-consuming than original cataloguing. In addition, for original cataloguing, the obtained average time is reduced to 9.26 minutes and the cost to €7.49. This represents about 21% less-time consuming and 29% less-costly than in the real case. Most strikingly, is the case of copy cataloguing in which the reduction is approximately 32% in time and 44% in cost. Therefore, the results of the what-if analysis confirm the validity of implementing these changes in the cataloguing process flow as part of a best practices approach. The enhanced data flow diagram of the cataloguing process is shown in Figure 5.

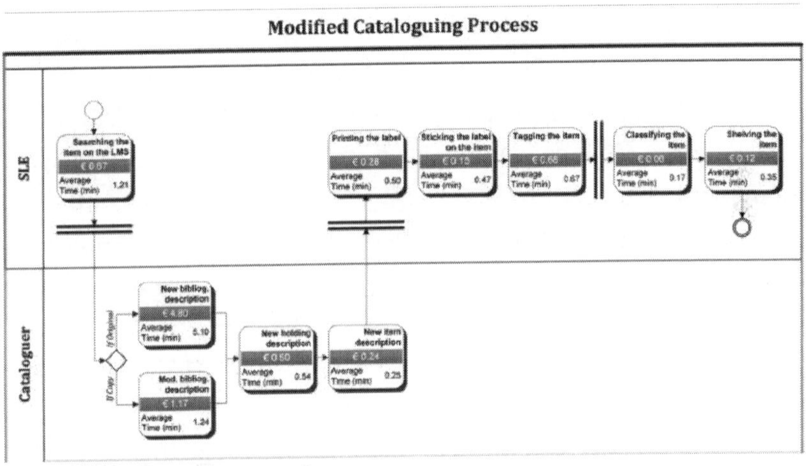

Note. – LMS = Library Management System.

Figure 5. Improvements on the cataloguing process.

5. BENEFITS OF TDABC IN CATALOGUING PROCESSES

The implementation of TDABC in cataloguing processes at CBA demonstrated important benefits. The first important benefit, shows the possibility to clearly *discriminate activities regarding the time*, and thus determining which activities demand more time and cost. For instance, analysis showed that labelling and shelving activities consume unnecessary resources. The second benefit is a consequence of the previous one. That is, the possibility of performing "what-if analysis" to *simulate potential scenarios;* for example, as labelling and shelving consumed unnecessary resources, what-if analyses were conducted. The first simulation showed that by delegating these activities to SLE staff, the library manager could reduce up to 23% of costs. The second simulation improves the previous simulation by incorporating batching processes. Results showed a reduction of approximately 37% of cataloguing costs, allowing librarians to improve their processes and liberate their time to perform other specialized activities. An additional benefit is that TDABC allows *benchmarking different scenarios locally and among libraries*, for example, original versus copy cataloguing. These results show that original cataloguing in this particular scenario is 30% more time-consuming, and consequently more costly than original cataloguing. Moreover, by benchmarking time and cost of activities among libraries, TDABC allows libraries to adopt policies and procedures to improve efficiency. When sharing benchmark figures, Slight-Gibney (1999) states that for other libraries, the time spent on various activities is probably more useful than the costs. In the case of benchmarking cost, the participating libraries require a common understanding of how to attribute indirect costs to their calculations and of what costs are included/excluded in order to have standard results. This problem does not occur when benchmarking the time spent on different activities. Eventually, a fourth important benefit is that TDABC allows *justifying decisions and choices.* TDABC allows both managers and staff to better understand alternative options and accept the need for change; for example, the decision of transferring responsibilities from cataloguers to SLE staff, as well as, structuring batch activities (processing, labelling and shelving). In fact, library managers should constantly analyse their cost information and keep their models updated in order to redesign workflows efficiently and effectively, as well as to reallocate resources and tasks.

Nevertheless, **Siguenza-Guzman et al. (2013b, 2014)** suggest important considerations to be borne in mind at implementing TDABC in academic libraries, namely: 1) the resource intensity of data collection to gather the time duration of activities, as well as to document the activity flows; 2) the need of a dedicated software tool to keep the flows updated and consequently to facilitate long-term maintenance; and 3) the commitment of library managers and staff during the data collection.

6. CONCLUSION

In this paper, the TDABC implementation was described in two main cataloguing activities, namely, original and copy cataloguing. These processes were selected because they are considered to be a part of the core activities of a library to manage its collection, but are also resource intensive. Based on our findings, we can conclude that TDABC is a quicker and easier way of calculating cataloguing savings. In fact, TDABC is a useful method to perform cost analysis in cataloguing processes, and consequently provides valuable data for managerial decisions. The TDABC implementation provided library managers with important information about cataloguing costs and performance measurements; and guided decisions concerning resource allocation and process improvements. For example, based on the obtained results, the library manager decided to delegate certain activities and define a set of batch activities. This case study, therefore, shows significant contributions to the literature on the implementation of advanced cost models for library processes, and more precisely for cataloguing activities.

A potential direction for future research is to expand this study to different cataloguing activities such as cooperative, contract, and outsourcing cataloguing. TDABC can also offer the possibility to discuss how these trends in cataloguing processes affect cataloguing units. TDABC will allow to examine whether these trends really provide an opportunity for cataloguers to spend their time on other cataloguing activities, such as enhancing existing records, or not. Another interesting area for future research would be a similar analysis for cataloguing audio-visual items and other special materials such as old books, non-print materials, and maps. Utilizing TDABC to benchmark libraries for "best practices" is an additional prospect for future analysis.

NOTES

[1]http://bib.kuleuven.be/cba/

ACKNOWLEDGMENTS

We wish to thank the staff of the Arenberg Campus Library at KU Leuven for their cooperation throughout this study; Hilde Van Kiel, head of the Library, for her big support and guidance; and Marc Verbrugge for his assistance during the data collection phase. The authors also want to thank Paul Vanegas for his comments on the paper. This research project was funded by the University of Cuenca, the Flemish Interuniversity Council (VLIR-IUC), and the National Secretariat of Higher Education, Science, Technology and Innovation of Ecuador (SENESCYT).

REFERENCES

1. ACRL Research Planning and Review Committee. (2010). 2010 top ten trends in academic libraries: A review of the current literature. College & Research Libraries News, 71(6), 286–292

2. Beall, J., & Kafadar, K. (2004). The effectiveness of copy cataloging at eliminating typographical errors in shared

bibliographic records. Library Resources & Technical Services, 48(2), 92–101

3. Campus Bibliotheek Arenberg. (2013, December 23). CBA - Campusbibliotheek Arenberg. Retrieved January 22, 2014,

4. Charbonneau, M. D. (2005). Production benchmarks for catalogers in academic libraries. Library Resources & Technical Services, 49(1), 40–48

5. Ching, S. H., Leung, M. W., Fidow, M., & Huang, K. L. (2008). Allocating costs in the business operation of library consortium: The case study of Super e-Book Consortium. Library Collections, Acquisitions, and Technical Services, 32(2), 97–103.

6. Cooper, R., & Kaplan, R. S. (1988). Measure costs right: Make the right decision. Harvard Business Review, 66(5), 96–103

7. Cottrell, T. (Terry). (2012). Three phantom budget cuts and how to avoid them. Bottom Line: Managing Library Finances, The, 25(1), 16–20.

8. Ellis-Newman, J. (2003). Activity-Based Costing in user services of an academic library. Library Trends, 51(3), 333–348

9. Ellis-Newman, J., Izan, H., & Robinson, P. (1996). Costing support services in universities: an application of activity-based costing. Journal of Institutional Research in Australasia, 5(1), 75–86

10. Ellis-Newman, J., & Robinson, P. (1998). The cost of library services: Activity-based costing in an Australian academic

library. The Journal of Academic Librarianship, 24(5), 373–379.

11. Everaert, P., Bruggeman, W., Sarens, G., Anderson, S. R., & Levant, Y. (2008). Cost modeling in logistics using time-driven ABC: Experiences from a wholesaler. International Journal of Physical Distribution & Logistics Management, 38(3), 172–191.

12. Goddard, A., & Ooi, K. (1998). Activity-Based Costing and central overhead cost allocation in universities: A case study. Public Money and Management, 18(3), 31–38.

13. Hoozée, S., Vermeire, L., & Bruggeman, W. (2012). The impact of refinement on the accuracy of Time-Driven ABC. Abacus, 48(4), 439–472.

14. Kaplan, R. S., & Anderson, S. R. (2004). Time-Driven Activity-Based Costing-Tool Kit. Harvard Business Review, (82), 131–138

15. Kaplan, R. S., & Anderson, S. R. (2007). Time-Driven Activity-Based Costing: A simpler and more powerful path to higher profits. Boston, MA, USA: Harvard Business School Press

16. Kaplan, R. S., & Cooper, R. (1998). Cost & Effect: Using integrated cost systems to drive profitability and performance. Boston, MA, USA: Harvard Business School Press

17. Kont, K.-R. (2011). New cost accounting models in measuring of library employees' performance. Library Management, 33(1/2), 50–65.

18. Manaf, Z. A., & Rahman, R. A. (2006). Examining the quality of National Library of Malaysia (NLM) cataloguing in

publication (CIP) records. Library Review, 55(6), 363-373.

19. McCain, C., & Shorten, J. (2002). Cataloging efficiency and effectiveness. Library Resources & Technical Services, 46(1), 23–31. doi:10.5860/lrts.46n1.23

20. Mitchell, A. M., Thompson, J. M., & Wu, A. (2010). Agile cataloging: Staffing and skills for a bibliographic future. Cataloging & Classification Quarterly, 48(6-7), 506–524.

21. Morris, D. E. (1992). Staff time and costs for cataloging. Library Resources & Technical Services, 36(1), 79–95

22. Morris, D. E., Hobert, C. B., Osmus, L., & Wool, G. (2000). Cataloging staff costs revisited. Library Resources & Technical Services, 44(2), 70–83

23. Morris, D. E., & Wool, G. (1999). Cataloging: Librarianship's best bargain. Library Journal, 124(11), 44-46

24. Novak, D. D., Paulos, A., & Clair, G. S. (2011). Data-driven budget reductions: A case study. The Bottom Line: Managing Library Finances, 24(1), 24–34.

25. Pernot, E., Roodhooft, F., & Van den Abbeele, A. (2007). Time-Driven Activity-Based Costing for inter-library services: A case study in a university. The Journal of Academic Librarianship, 33(5), 551–560.

26. Reitz, J. M. (2004). ABC-CLIO - ODLIS. ODLIS - Online dictionary for library and information science. Libraries Unlimited. Retrieved December 18, 2013, from http://www.abc-clio.com/odlis/odlis_a.aspx

27. Roberts, S. A. (2003). Financial management of libraries: Past trends and future prospects. Library Trends, 51(3), 462–493

28. Siguenza-Guzman, L., Van den Abbeele, A., Vandewalle, J., Verhaaren, H., & Cattrysse, D. (2013a). Recent evolutions in costing systems: A literature review of Time-Driven Activity-Based Costing. ReBEL - Review of Business and Economic Literature, 58(1), 34–64

29. Siguenza-Guzman, L., Van den Abbeele, A., Vandewalle, J., Verhaaren, H., & Cattrysse, D. (2013b). Towards a holistic analysis tool to support decision-making in libraries. In Proceedings of the IATUL Conferences. Cape Town, South Africa: Purdue e-Pubs. Retrieved December 18, 2013, from http://docs.lib.purdue.edu/iatul/2013/papers/29/

30. Siguenza-Guzman, L., Van den Abbeele, A., Vandewalle, J., Verhaaren, H., & Cattrysse, D. (2014). Using Time-Driven Activity-Based Costing to support library management decisions: A case study for lending and returning processes. Library Quarterly: Information, Community, Policy, 84(1), 76–98.

31. Slight–Gibney, N. (1999). How far have we come? Benchmarking time and costs for monograph purchasing. Library Collections, Acquisitions, and Technical Services, 23(1), 47–59.

32. Stouthuysen, K., Swiggers, M., Reheul, A.-M., & Roodhooft, F. (2010). Time-Driven Activity-Based Costing for a library acquisition process: A case study in a Belgian University. Library Collections, Acquisitions, and Technical Services, 34(2–3), 83–91.

33. Vazakidis, A., & Karagiannis, I. (2009). Activity-Based Management and traditional costing in tourist enterprises (A hotel implementation model). Operational Research, 11(2), 123–147.

34. Wegmann, G., & Nozile, S. (2009). The Activity-Based Costing method developments: State-of-the art and case study. The IUP Journal of Accounting Research and Audit Practices, 8(1), 7–22. doi:oai:halshs.archives-ouvertes.fr:halshs-00482137

35. Yilmaz, R. (2008). Creating the profit focused organization using Time-Driven Activity Based Costing. In EABR & TLC Conferences Proceedings (p. 8). Salzburg, Austria: Clute Institute for Academic Research

4

CHAPTER

The Changing Role of the Subject Specialist

Michael Cotta-Schønberg

ABSTRACT

As we all know, libraries are these years rapidly undergoing change on unparalleled scale. Evidently, this applies to librarians, too, and not the least to that important category of library staff, the subject specialist. As recruiting and education of library workers differ from country to country it is difficult to give a detailed, generally valid description of the subject librarian in libraries, but I believe that you can describe an ideal model of subject librarianship as follows: Within each of the major subject disciplines covered by the library, the library should have a subject specialist preferably with a master degree or at least a bachelor degree in the particular subject discipline. The role of the subject specialist is to perform four basic functions where extensive subject knowledge is considered to be necessary: selecting and classifying books, assisting users with advanced subject inquiries, giving subject-specific courses in information retrieval, and maintaining liaison with relevant academic departments and centres. Personally, I know this system very well since I got employment in the Royal Library in Copenhagen as a subject specialist in psychology in the very month I finished my degree in psychology from the University of Copenhagen, back in 1973. The subject librarian system at the Royal Library in Copenhagen was patterned on the ideal model, as I just described it, and it was closely paralleled in the other academic libraries in Denmark, also the new university libraries which were founded in the seventies.

THE IDEAL MODEL

As we all know, libraries are these years rapidly undergoing change on unparalleled scale. Evidently, this applies to librarians, too, and not the least to that important category of library staff, the subject specialist. As recruiting and education of library workers differ from country to country it is difficult to give a detailed, generally valid description of the subject librarian in libraries, but I believe that you can describe an ideal model of subject librarianship as follows:

Within each of the major subject disciplines covered by the library, the library should have a subject specialist preferably with a master degree or at least a bachelor degree in the particular subject discipline. The role of the subject specialist is to perform four basic functions where extensive subject knowledge is considered to be necessary: selecting and classifying books, assisting users with advanced subject inquiries, giving subject-specific courses in information retrieval, and maintaining liaison with relevant academic departments and centres.

Personally, I know this system very well since I got employment in the Royal Library in Copenhagen as a subject specialist in psychology in the very month I finished my degree in psychology from the University of Copenhagen, back in 1973. The subject librarian system at the Royal Library in Copenhagen was patterned on the ideal model, as I just described it, and it was closely paralleled in the other academic libraries in Denmark, also the new university libraries which were founded in the seventies.

Problems of the ideal model

During my career I have seen a number of cases where the ideal model functioned very well, due to the dedication and skills of individual subject specialists. Today, however, the ideal model has a number of problems.

The financial problem

The first and basic problem is - as usual - financial. The Copenhagen University Library together with the National Library comprises the Royal Library in Copenhagen. It covers all scholarly disciplines within the humanities, social science, theology, law, natural science, medicine and dentistry, agriculture and pharmaceutical science. The ideal model would require that we should employ at least fifty university graduates with master degrees in the various subject fields covered by the university. This is definitely not possible. That staff category is the best salaried in the library, and we simply cannot afford fifty or more staff members in that category.

The financial problem was always there, and in fact the library had to make many compromises with the ideal model in the sense that people were asked to function as subject specialists in areas in which they had no university education. Such arrangements have actually functioned very well, but because of the persistence of the ideal model as a mental norm, they are looked upon as improper arrangements, especially by proper subject specialists: "proper" subject specialists feel that their primary function as well as the quality of library services is devalued by library management, when it appoints subject specialists without university degrees in the particular subject. And such subject specialist may feel insecure, uncertain and inadequate in the exercise of their function.

Relations with other staff

Secondly, the presence of a highly educated group with university degrees on master level tends to create an A-team and a B-team within the workforce of the library. The subject specialists are of course the A-team. When I was a young subject specialist at the Royal Library, we subject specialists were the barons of the system, each one in his own subject barony. The fact that the university-educated subject specialists monopolise the very important subject-related functions in the library is unavoidable in a library structure based on subject specialisation.

And there clearly is a lower limit to what other library tasks you can use those university graduates for, they tend to attract a disproportionate part of other, high-level non-subject related functions or projects of the library. This means that other staff groups, especially the library school educated librarians, tend to become less involved in development or other interesting projects, which means that the potential of that group is not used optimally. The situation is not a healthy one in terms of organizational climate and it does not promote the best use of staff resources.

Status vis-à-vis faculty

Paradoxically, the high status of university-educated subject librarians in the library itself is not at all reflected in its status vis-à-vis faculty. Whereas the library's subject specialists should be - and admittedly in many cases are - the privileged medium of contact between library and faculty, quite often the subject specialists tend to remain within the library sphere and do not have those lively partnerships with faculty, which are essential to the development of library service to universities.

It does seem that subject library staff does not generally have high status with university faculty. To research and teaching staff, libraries are infrastructure and subjects specialists are part of that infrastructure, they are service personnel. This perception may be coloured by the fact that university graduates (outside the library discipline), who end up in libraries, generally do not belong to and are not perceived by faculty to belong to the cream of the academic crop, which will in stead - of course - try to obtain tenure at the university. If this is correct, it will of course tend to keep the two parties, faculty and subject specialists, apart and it will not favour the role of university-educated subject specialists as the vital link between library and faculty.

The cross-disciplinary university

A fourth problem concerns subject specialisation in itself. In the modern university there is a trend towards cross-disciplinary research and development, and the traditional division of knowledge tends to give way to other, dynamic subject profiles. As the subject structure of the university library generally reflects the traditional divisions of knowledge, it increasingly gets out of sync with the dynamic subject profile and departmental structure of the university. This development tends to favour the development of another type of subject specialist with less emphasis on deep subject knowledge in individual subject specialists and more emphasis on varied, combinatorial and cross-disciplinary knowledge profiles.

New forms of book selection and subject classification

In former days, book selection by subject specialists was to a large extent based on reviews in learned periodicals or on sources with little information on the books. Book selection therefore required intensive subject knowledge. The same applied to the subject data selected for each book by the subject specialist, often based on local classification systems.

Today, the book selector has access to a wide array of excellent digital sources for book selection: new records in national bibliographies or book dealers' online selection bases, often augmented with tables of contents and abstracts and with access reader's reviews and I know not what. Moreover, the move towards digital resources means that books will to a larger extent be bought not on a title-by-title basis but as general packages, for example from publishers or from aggregator systems like e-library.

As for subject classification or indexing more and more libraries are relying heavily on data imported from external sources, instead of producing them locally. The net result is that the need for staff with deep subject knowledge to select books and provide subject data has lessened considerably.

The changing nature of reference work and user instruction

It seems to be a general phenomenon that work at libraries' reference desks is changing from the more complicated subject enquiries to practical, technical and ICT-related problems and to more general questions about searching procedures. One reason may be that users themselves are becoming more and more proficient in handling information searching on the web and less dependent on librarians to formulate particular subject search strategies.

The same applies to user instruction. Formerly, there was a strong emphasis on instruction in the use of subject-specific reference works and abstracting and indexing sources. Today, the emphasis is more on the development of general information competence.

CONCLUSION

The combined effect of these various developments is that libraries' need for subject specialists with intensive subject knowledge is diminishing. University libraries will still have - and need - staff with university degrees in the subjects covered by the university. But their job profile will change, their traditional functions will arguably become less important and their place in the staff structure will be different.

The future: from subject specialist to information specialist

I believe that the general trends, I have described, will be felt much the same way in many traditional university libraries. As local conditions do vary greatly, there is no "ideal" solution to the problems. It may actually be an "evolutionary" advantage if different solutions are tried out so that all may eventually benefit from the best ones.

Now I'll turn to the solution we are implementing in the Copenhagen University Library, not as a course I recommend to all, but as an input to the discussion of an important organisational theme for university libraries today. For us, the two major issues to be addressed in this area are 1) service development and b) staffing structure and staff development. To know what kind of staff we need, we first have to know what kind of service we want to provide.

Our service will be based more and more on large packages of digital knowledge resources, comprising both databases, e-journals and e-books. To organise this service and deliver it to users will be the major task for our staff, and the traditional library functions based on print resources will no longer constitute the fundamental structuring principle for the organisation of our library.

Secondly, we have decided that direct personal service using all available channels will still be an important and integral part of our service. By all available channels I mean electronic, telephonic, postal, personal channels of every kind, but it is rapidly becoming quite clear that if we want to be where the users are, we have to develop a strong presence on the web.

We do have other strategic aims, but these two are the most important in terms of staffing structure and staff development.

What kind of staff do we need for our new university library?

1. Our staff will be specialists in the organisation and delivery of knowledge resources to the university

2. They will have broad subject knowledge allowing them to apply their information skills to the rapidly changing research and teaching profile of the university. They must not each have deep subject knowledge in individual disciplines, I am more interested in staff who have a general grounding in broader areas like humanities, social sciences, natural sciences etc.

3. Our staff must have highly developed social skills allowing them to interact naturally and competently with users. The time of back office staff is over. Apart from reasonable exceptions - and I mean exceptions – all staff will have mixed job profiles but with a heavy element of direct user contact.

In our context, these demands point to the recruiting of staff with combinatory educations and skills. The preferred job applicant will be someone with an education combining library and information science or IT and web with some university subject, preferably a broad one. Moreover, applicants with job experience from the service sector and who ooze sociability will be highly valued in the recruiting process.

But recruiting staff is not enough, we must also have staff development. You are all familiar, of course, with modern methods of staff development, so I shall not go deeply into that. I shall only point to what for me is the most important instrument of staff development. That is job rotation. All new staff will change jobs in the library every third year or more often, even. During their career in our library they shall be moving across functions, subjects, departments. Not only will they acquire extensive knowledge of the whole library and the whole university, but moving along they will be developing their personal networks. This networking is very important in an organisation as large as ours and moreover scattered all over Copenhagen.

We have now established a new staff category of information specialists and have stopped recruiting staff into the categories of librarian (with a library education from the library school) and research librarian/subject specialist (with a degree in some university discipline). As of December 2006 all new staff apart from technical and administrative staff is recruited into this new category of information specialist. And all new staff (include clerical staff) rotate on a three year basis.

I have been asked by one of my bright young people why we call the new staff category information specialists, when what we are really aiming at is information generalists. The term information specialist defines the job in relation to other jobs outside the profession, like teachers, artists etc. Within the library it delimits the group in relation to technical and clerical staff.

The important thing is to signalize that the university library is developing into a new type of institution with diminishing emphasis on printed books and increasing emphasis on information in the sense of mediation of knowledge resources in all forms and along all channels. The same applies to staff: the term information specialist indicates a shift away from particular functions or subjects to information as the defining task of library staff.

Only time will show, of course, whether this is a good move or not. I myself believe in it firmly and I am encouraged by the fact that applicants for jobs in the library seem to prefer the concept of information specialist to the concept of librarian, and moreover they heartily endorse our new system for job rotation.

5

CHAPTER

Leveraging the Power of a Twitter Network for Library Promotion

Jason Shulman[a,], Jewelry Yep[b], Daniel Tomé[c]

[a] *Department of Physics, Richard Stockton College of New Jersey, 101 Vera King Farris Dr., Galloway, NJ 08205, USA*

[b] *School of Health Sciences, Richard Stockton College of New Jersey, 101 Vera King Farris Dr., Galloway, NJ 08205, USA*

[c] *Office of Service Learning, Richard Stockton College of New Jersey, 101 Vera King Farris Dr., Galloway, NJ 08205, USA*

ABSTRACT

The Twitter network of two academic libraries was analyzed to determine the influential accounts that connect to them. Such information can be exploited by libraries to create tailored social media outreach and information dissemination programs. Three network metrics, measuring different definitions of importance, were calculated for each account in the network. This allowed for the quantification and ranking of the accounts by influence/importance, normally considered to be qualitative and subjective. By all measures, accounts associated with the institutions, and not faculty, staff, or students, were found to be the most influential players in the networks of both libraries, suggesting that this is a general feature of academic library Twitter networks. Furthermore, the library, as an institutional account itself, is also influential to the broader Twitter community of its home institution. This demonstrates that the library is in a key position to propagate information from sister accounts at the institution.

Keywords: Twitter; Social media; Networks; Influence; Metrics; Follower

INTRODUCTION

Academic libraries, in an effort to provide convenient and effective service, have been quick to implement new technologies, evaluate their efficacy, and refine their use. Perhaps the most significant recent example of this has been the incorporation of social networking into promotion and outreach activities. In particular, Twitter has become one of the most widely adopted and studied platforms. Libraries have used Twitter primarily for marketing of services and programs (Del Bosque et al., 2012, Gunton and Davis, 2012 and Milstein, 2009). Many aspects of a library's Twitter program affect its effectiveness; however, they can be divided into two primary categories, properties of the tweets and properties of the associated accounts (Petrovic et al., 2011, Suh et al., 2010 and Yang and Counts, 2010). Milstein (2009) and Cole (2009) present best practices involving the former. This work focuses on the latter.

Information dissemination is the primary goal of any Twitter program run by the library. In analyzing the success of such endeavors, we often simply look to the number of followers of the account. It is certainly true that the information reaches more people as the number of followers increases; however, relying solely on the number of followers to gauge the impact of a Twitter program ignores much of what makes social media an effective vehicle for outreach and dissemination of information, the network. Indeed, Twitter accounts form a network. Accounts follow and/or are followed by the library, but these accounts can follow one another and even others not directly connected to the library. Links between accounts that are not directly connected to the library can have significant impact on information dissemination. If properly harnessed, these links can help spread the library's message well beyond its direct followers (Yep & Shulman, 2014).

This work presents an analysis of the follower/followee networks of the libraries from two primarily undergraduate state institutions, The Richard Stockton College of New Jersey and California State University San Marcos, with the goal of identifying the influential accounts connected to the library. Such information can allow those in charge of a library's account to tailor their tweeting and increase the reach and effectiveness of Twitter activities. Three metrics related to an account's influence were examined. Interestingly, it was found that the most influential accounts, by all measures, were not the students nor the faculty, but the other accounts associated with the institutions. Similar results from the analyses of both schools suggest that this is a general feature of such networks. The consequences of this finding will be detailed below. Importantly, the procedure employed in this study can be easily implemented by librarians to identify specific influential accounts at their institutions, detect communities within the network, and tailor their Twitter activities to maximize information dissemination.

A story reported by Harold Glazer (2009) of Rutgers University nicely illustrates how a single social media connection can enhance a library's electronic outreach efforts. Glazer describes the implementation of Facebook at his library. Early in the program, he noticed a marked increase in library related articles in the school's student newspaper, several of which were featured on the front page. Glazer traced the source of this new attention to the editor of the paper, who had connected to Glazer and the library on Facebook. The moral of this story, for the present work, is that not all social media connections are equal. As much as we might not like to admit it, some accounts wield more influence than others. The purpose of this study is to identify the influential players in the Twitter network so that they may be recruited for dissemination purposes. It also allows for the quantification of an account's influence, something that is normally qualitative. Finally, the procedure outlined below can be automated by a computer and requires few resources from library staff.

LITERATURE REVIEW

Social media use by patrons and libraries

Two comprehensive reports were presented by Cassidy et al., 2011 and Cassidy et al., 2014 on student technology use at Sam Houston State University. The most recent study found that more than 80% of students had a smartphone, which has led to a larger demand for mobile library services. The reports also show that student adoption of Twitter has doubled, from 21.2% to 41.4%, during the intervening years. Sixty percent of students in the 2014 study were not interested in library services using Twitter; however, the majority of Twitter users did indicate desire to connect with the library *via* Twitter.

Park (2010) examined the use of the popular Korean social networking site Cyworld by students, both undergraduate and graduate, and faculty. He found undergraduates open to adopting new technologies and interested in exposure and ability for self-expression afforded to them by Cyworld. In contrast, graduate students and faculty preferred to use the internet for information gathering purposes. Faculty demonstrated little social media use, while graduate students employed it for assistance with schoolwork and career advancement. Park concludes that efforts by libraries to engage faculty *via* social media should highlight the benefits of its use to social networking and communication.

In fact, Dickson and Holley (2010) suggest that academic libraries must advertise their social media services. As mentioned previously, Rutgers's library benefitted from the exposure their Facebook page received in the student newspaper. Perhaps one of the more interesting examples of promotion is the use of a Chinese microblogging site to successfully promote a social media marketing campaign (Luo, Wang, & Han, 2013). It's worth noting that the procedure presented below can help librarians market the account and their services by identifying influential Twitter accounts which can be employed to increase the library's exposure.

Twitter use by libraries

Twitter is most commonly used to broadcast information about the library (Gunton and Davis, 2012 and Milstein, 2009). Eleven out of twenty suggestions provided by Cole (2009) involve sending out information to followers, while five are devoted to promotion of the library and its services. In their broad review of Twitter use by academic libraries, Del Bosque et al. (2012) found that libraries primarily used their accounts for discussing resources (55% of libraries), announcing events (24%), and communicating hours (14%).

While Twitter is an effective tool for broadcasting library information, the scholarly community appears to be united in the belief that Twitter should be used by academic libraries to interact with followers in order to take full advantage of the platform (Cole, 2009, Cuddy et al., 2010, Del Bosque et al.,

2012, Dickson and Holley, 2010, Gunton and Davis, 2012, Milstein, 2009 and Sewell, 2013). Gunton and Davis (2012) suggest that limiting a library's social media activities to the distribution of information represents a failure to appreciate the interactive nature of the modern internet. It appears that academic libraries have been slow to adopt such advice. Del Bosque et al. (2012) found that 54% of libraries interacted with followers and less than 10% used Twitter to answer reference questions, although most libraries did have some version of electronic reference. The authors encourage libraries to take advantage of Twitter-specific features such as direct messages, @replies (replying to an account by using @ followed by the account name), and hashtags (keywords preceded by the # symbol, *e.g.*#LibraryScience). Hashtags allow Twitter users to easily follow discussions on a topic simply by searching for the hashtag. Del Bosque et al. also advocate the use of library-specific hashtags.

Information dissemination *via* Twitter

It is clear that social media is used by academic libraries as a means to transmit information to and connect with patrons. Thus, an effective program will reach many people. This is a question of information dissemination. Dissemination *via* Twitter has been studied by several groups. Much of the focus has been on retweeting, or the forwarding of a tweet composed by another, since, according to Suh et al. (2010), "Retweeting is the key mechanism for information diffusion in Twitter." Several account properties are associated with high levels of retweeting, the most obvious of which is the number of followers (Petrovic et al., 2011 and Suh et al., 2010). Perhaps more surprisingly, the number of followees is also related to the generation of retweets (Suh et al., 2010) as is the account being a member of Twitter lists (Petrovic et al., 2011).

Kim, Abels, and Yang (2012) studied information dissemination by academic libraries. The study examined the account types that retweeted library content. Retweeters were grouped into twelve categories, *e.g.* librarians, students, scholars, and university organizations. University organizations did the most retweeting and was the largest intermediary, garnering more retweets of library messages than any other group. Such accounts are central to the Twitter network and are responsible for circulating much library content. Students were found to retweet messages both directly from the library and also those that were previously retweeted by other groups. A moderate amount of retweeters were categorized as local organizations; however, this group was responsible for spreading the second highest number of messages. This indicates the importance of cultivating relationships between academic libraries and such accounts.

This work is concerned with the Twitter networks and influential accounts within them. However, for completeness, it is worth noting properties of individual tweets that are associated with retweets and information propagation. The inclusion of both URLs and hashtags has been shown to increase the retweetability of a message (Petrovic et al., 2011, Suh et al., 2010 and Yang and Counts, 2010). Suh et al. (2010) found that the impact of including URLs in a tweet is domain specific. Some URLs enhance the appeal of the message while others reduce it. This was also found for hashtags. Overall, however, messages with these features are more likely to be forwarded by other users.

Analysis of followers

Sewell (2013) performed a comprehensive investigation of the Twitter followers from a library at Texas A&M University. Such an analysis is critical to the development of targeted marketing and relevant tweets (Cuddy et al., 2010). For example, this knowledge can indicate which communities are heavily represented, and moderators can tailor content to these groups. Furthermore, once known, underrepresented populations can be engaged and recruited.

There were 432 accounts following the Texas A&M library's Twitter account. Each was individually examined and placed into one of eleven categories, *e.g.* student, faculty/staff, and alumni. Other social media sites were examined to locate this information, if necessary. In addition, Sewell created other account properties such as affiliated/unaffiliated with the University, active/inactive, and number of tweets. Categories also had subproperties associated with them. For example, student accounts were further partitioned by academic program and year.

Forty-five percent of the followers were associated with the University. Students represented the largest population of followers (24%). Of these, 81% were undergraduates. Corporations (20%) were the second most represented population. Other groups to note were University departments/organizations (9%) and faculty/staff (5%). Cuddy et al. (2010) found a similarly diverse set of followers in the early days of the NYU Health Sciences Libraries' Twitter account.

As mentioned above, an analysis of an account's followers can provide valuable insight into the accounts that receive library information. It can help librarians determine policies ranging from simple tweet development to large-scale policy decisions such as whether to allocate resources to social media (Sewell, 2013). However, there are difficulties associated with such endeavors, the foremost of which is the enormous time investment. The account of each follower must be visited and details copied. Personal information, such as graduating year, department, or even gender is not always easily accessible. The account must be combed if such information is needed, requiring more time. Sewell was forced to search for and access other social media accounts to locate the information required by her study. Another difficulty with a follower examination is the analysis of such a large volume of irregular data. Developing a big picture from the many details of accounts is not straightforward. The purpose of this work is to complement this microscopic analysis of a library's Twitter community with a multiscale examination. A simple method to extract communities as well as key accounts in the Twitter network is presented. Importantly, the data collection and analysis is automated, thus requiring little time from librarians.

INTRODUCTION TO NETWORKS

The field of networks is vast and continues to grow. This section will introduce only that which is relevant to the present study. More details can be found in the works of Newman (2010) and Easley and Kleinberg (2010).

Terminology

A network is a collection of entities that connect or interact in some manner. This definition is quite general, which allows the concept to be applied to diverse situations. Consider the following examples. In a friendship network, the entities are people and the connections are friendships. The power grid is also a network. The entities are power plants, businesses and homes, and the connections are power lines. Citation networks are formed from articles and books that cite other written works.

These examples demonstrate that the concept of networks has permeated many fields. As a consequence, redundant terminology has developed. The entities in a network are commonly referred to as *vertices* but are also known as *nodes*. The connections are called *edges*, *interactions*, or *ties*, to name a few. Networks can be visualized by drawing a network *graph* (see Fig. 1). Vertices are represented as dots (or similar) and edges are drawn as lines connecting them. In some networks, a direction can be associated with the edges. For example, in a follower/followee Twitter network, an account can follow another, but that does not mean that the attention is reciprocated. Fig. 1 shows an example of such network. Alice follows Bob, as indicated by an arrow running from her to Bob. An arrow also points from Bob to Alice, demonstrating that he follows her too. Claire follows Dan; however, the interaction is not reciprocated. Networks in which edges have a direction are referred to as *directed networks*. In a follower/followee Twitter network, attention follows the direction of the arrows. Claire pays attention to Dan's tweets. Bob pays attention to Claire, while Alice pays attention to Bob who, in turn, pays attention to her. Note, however, that information flows *against* the arrows, in the opposite direction of attention.

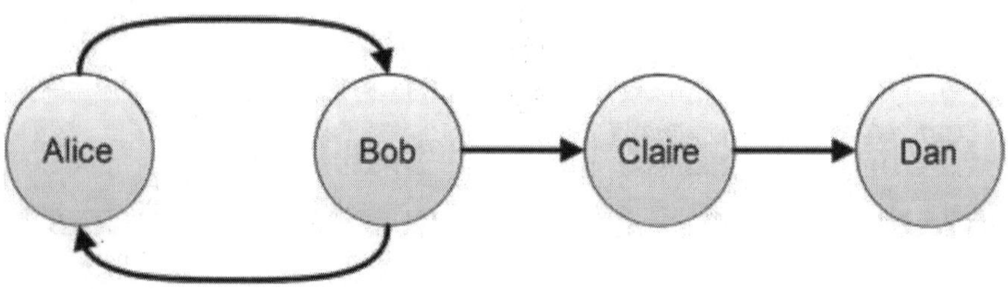

Figure 1. A sample follower/followee Twitter network graph. Arrows' direction indicates attention paid by one account to another. They begin and end on accounts that follow and are followed, respectively. Information, in the form of tweets, flows against the arrow direction.

Metrics

One might imagine that vertices in a Twitter network can possess differing levels of influence; some are more important in the network than others. Celebrities, for example, are retweeted much more often than others, making them sources for information and allowing them to wield considerable influence (Petrovic et al., 2011, p. 587). The importance of a vertex can be defined in many ways. Perhaps it is popular, possessing many connections. Vertices that act as a conduit for information can also be considered important, even if they do not have many connections. Sociologists have developed methods to identify central vertices in a network and rank them according to these different definitions of importance. The measures of importance are called *centralities*. This work discusses a few of them.

Degree

Degree, sometimes called degree centrality, is the number of edges attached to a vertex. In a friendship network, for example, a person's degree would be his or her total number of friends. Thus, degree can be considered a rudimentary measure of popularity (Hansen, Shneiderman, & Smith, 2010, p. 40). For directed networks, such as the Twitter follower/followee considered here, there are two degree measures, in-degree and out-degree, describing the incoming and outgoing connections, respectively. The in-degree is a measure of popularity, a measure of the number of followers within the network. The out-degree is a measure of how much attention an account pays to the others in the network. In Fig. 1, Alice has an in-degree of one and an out-degree of one. Dan has an in-degree of one and an out-degree of zero; Claire follows him, but he does not follow anyone in the network.

Eigenvector centrality

There is a limitation to using a vertex's degree to quantify its significance in the network. Each connection is valued equally, so forming a connection with an important vertex counts as much as a connection to an unimportant one. It can be argued, however, that developing a connection with the head of one's company bestows more influence than a connection with someone in an entry level position. Eigenvector centrality[1] accounts for this discrepancy. Like degree, the value of the centrality can be large by having many connections. It can also be large if the vertex has important neighbors (Newman, 2010). In a social network, a person with many friends is influential. Eigenvector centrality accounts for the fact that a person can be influential with only a few friends who happen to be influential themselves.

Betweenness centrality

Imagine a piece of information flowing between two accounts in a Twitter network. Likely, there will be multiple paths that connect the two accounts. Consider two paths, one long and one short. The first account generates the information with a tweet, and those following the account have the option to retweet it. In fact, each account that receives the information along the chain has the option to pass it along or simply ignore it. If the path is long, with many accounts, the likelihood that all accounts will pass along the information is low; often someone will ignore it, and the information flow will die en route to the account at the end of the chain. Tweets rarely make it five steps away from the source account (Yang & Counts, 2010, p. 357). Thus, short paths tend to be important in passing along information and, consequently, accounts that lie on many short paths have considerable control over information diffusion in the network. This property is captured in a metric called *betweenness centrality*. Vertices with a high betweenness centrality lie on many of the shortest paths between the other vertices in the network. They act as gatekeepers of information.

Fig. 2 shows a network in which each vertex is decorated with its degree (below) and betweenness centrality (above). Vertex E acts as a bridge connected the clusters on the left and right side of the graph. It has the highest value of betweenness centrality because any information flowing from the left cluster to the right (and *vice versa*) must pass through vertex E. In fact, if E was removed from the network, there could be no communication between the left and the right sides. Clearly vertex E is important to the network. Note, however, that it has the lowest degree (a tie). This example underscores the need to examine several metrics. By considering only the degree of the vertices, one would rule out E as a significant player in the network. Vertices A, E, and I have the same degree; however, A and I have the lowest betweenness centrality. Unlike E, their removal would not impede information flow between the remaining vertices. Information could start at any vertex and reach any other, even if one or both are removed.

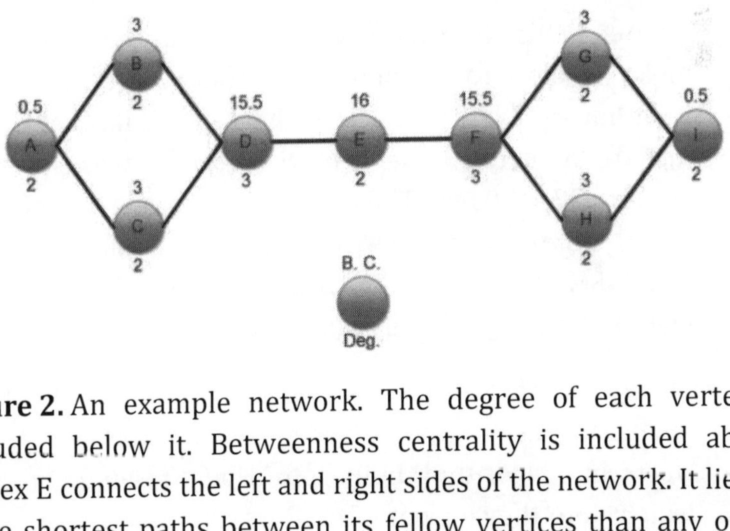

Figure 2. An example network. The degree of each vertex is included below it. Betweenness centrality is included above. Vertex E connects the left and right sides of the network. It lies on more shortest paths between its fellow vertices than any other.

Consequently, it has the largest betweenness centrality. It does not, however, have the largest degree.

METHODOLOGY

NodeXL[2] was used to both construct and analyze the follower/followee Twitter networks of the libraries. It was chosen for several reasons. First, it is a freely available add-on to the ubiquitous Microsoft Excel spreadsheet software. As such, familiarity with Excel translates into proficiency with NodeXL. More importantly, however, is NodeXL's automatic download of Twitter, and other social media, data. It also has built in network analysis tools. Therefore, access to and analysis of social media data can be accomplished in a matter of a few clicks of a mouse without requiring any programming knowledge from the user. This permitted the simple workflow used in this study, which is summarized below.

1. *Download data.* NodeXL was used to download the libraries' follower/followee networks from Twitter. Twitter limits the rate at which data can be accessed. It took 1–2 days to download the complete networks for each library. Data from the Richard Bjork Library at The Richard Stockton College of New Jersey was saved on March 27, 2014. The library's Twitter handle is @RSC_Library. The network of the Kellogg Library (@CSUSM_Library) of the California State University San Marcos was saved on July 20, 2014.

2. *Identify clusters.* Clusters within the networks were identified using the Clauset–Newman–Moore (CNM) clustering algorithm built into NodeXL. Fig. 2 can offer insight into the concept of clusters within a network. One might be tempted to partition the network into two

groups, vertices A–D and F–I, connected by vertex E. These two groups could be considered clusters of vertices within the network, although placement of vertex E is not obvious. Application of the CNM algorithm identifies two clusters, A–D and E–I. In larger networks, visually partitioning the network into clusters is not usually possible and algorithms and computers are required.

3. *Calculation of metrics.* NodeXL was used to calculate the in- and out-degree, eigenvector centrality and betweenness centrality of each vertex in the network. General metrics describing the network as a whole were also calculated, including the average and median in/out degree for the network as well as reciprocity, the likelihood that two connected vertices follow each other. These data were saved in the Excel file containing the network information.

4. *Visualization of networks.* NodeXL has powerful network visualization tools; however, we also used Gephi [3], another network analysis and visualization software package to produce some of the network graphs. In these cases, network data was exported from NodeXL into a GraphML file. This file was imported into Gephi.

RESULTS

General network properties

The follower/followee Twitter network for the library at The Richard Stockton College of New Jersey (RSC) is shown in Fig. 3. It has 205 vertices, *i.e.* 204 accounts follow or are followed by the library's Twitter account (or both). The network is constructed from 2485 edges, resulting in an average degree (both in and out) of 12.1. The median in-degree for the network is 3 while the median out-degree is 7. Reciprocity, *i.e.* the likelihood of finding two accounts that follow each other, was also measured for the RSC library network. A higher value indicates that there are more bidirectional edges, suggesting a strong relationship between the vertices (Hansen et al., 2010, p. 154). The library network has a reciprocity value of 28%. Finally, the CNM clustering algorithm identified four clusters, although one only contains two vertices. The first primary cluster includes the @RSC_Library account as well as accounts of professors, academic departments and other library related accounts. The other two primary clusters are dominated by accounts associated with the college, such as RSC's official Twitter account, and those associated with student organizations, such as the student senate. Personal student accounts are dispersed throughout the three primary clusters. See the Supplementary data for network graphs that segregate the clusters.

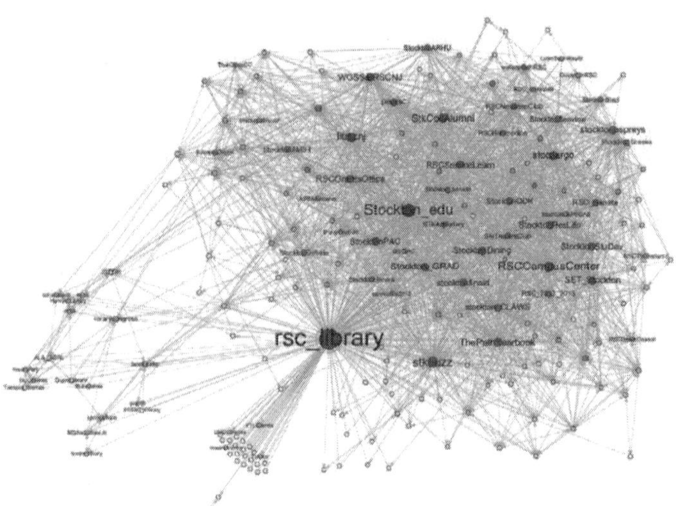

Figure 3. The RSC Library network. Vertex and label size are proportional to betweenness centrality. Vertex color reflects eigenvector centrality, with larger values corresponding to darker vertices. Labels for accounts owned by individuals are not included.

Interestingly, the network for the library at the California State University San Marcos (CSUSM) has values similar to those reported above for RSC's library (see Table 1). Five primary clusters were identified for the CSUSM network, however. An analysis of cluster membership did not identify specific populations inhabiting each cluster. Fig. 4 shows the CSUSM network.

Table 1. Network wide metrics for the RSC and CSUSM library networks.

Metric	RSC	CSUSM
# of vertices	205	244
# of edges	2485	2257
Mean in/out degree	12.1	9.3
Median in-degree	3	1
Median out-degree	7	6
Reciprocity	28%	23%
# of groups	4	5

Metrics

The values of in-degree, out-degree, eigenvector centrality and betweenness centrality were calculated for each vertex in the RSC and CSUSM networks. The vertices were then ranked according to the values of each metric. Table 2 and Table 3 show the ten most important vertices in the RSC and CSUSM networks, respectively, as determined by each metric. The library accounts themselves are not included in the table. They are, by definition, the most important vertices in their networks; they connect to all accounts. The purpose of the study is to identify the other accounts which

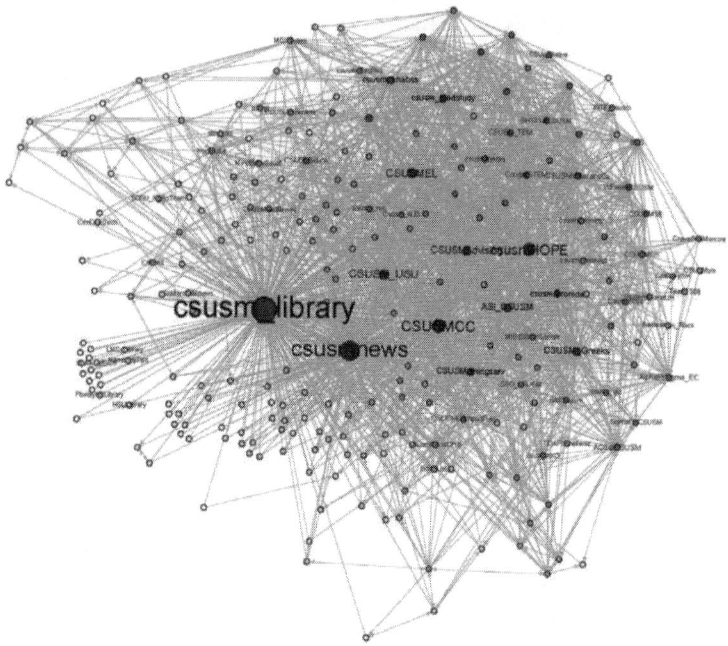

Figure 4. The CSUSM network. Same features as Fig. 3.

could be useful in disseminating library information. Out-degree is not included in the tables since it describes the number of accounts the vertex in question follows. Perhaps there are certain situations in which a large out-degree might imply importance, *e.g.* one might argue that a person who listens to many could be significant; however, out-degree matters little for situations concerning libraries and information dissemination. The values of the metrics are included next to the account names. The in-degree indicates the number of followers of that account. One is typically concerned with relative values of eigenvector and betweenness centralities. It should be noted that all of the accounts found in Table 2 and Table 3 are associated with the institution; there are no personal accounts of faculty, staff, or students. Each account is maintained by a division, a department, or a program at the institution. These are the types of accounts which are influential in academic library Twitter networks.

Table 2. RSC network accounts ranked in order of decreasing value of the metrics. The metric value is included in parentheses next to the account name. The median values are in parentheses next to their corresponding metric in the column title.

In-degree (3)	Eigenvector centrality (0.0031)	Betweenness centrality (2.4)
Stockton_edu (99)	Stockton_edu (0.0200)	Stockton_edu (2831)
RSCCampusCenter (87)	stkbuzz (0.0165)	stkbuzz (1222)
stocoargo (63)	litrscnj (0.0165)	RSCCampusCenter (1158)
StocktonPAC (62)	WGSSatRSCNJ (0.0165)	StkCollAlumni (660)
StocktonStuDev (60)	RSCServiceLearn (0.0161)	litrscnj (652)
StocktonDining (60)	RSCCampusCenter (0.0159)	stocoargo (480)
Stockton_GRAD (60)	peprsc (0.0158)	Stockton_GRAD (467)
StkCollAlumni (60)	Stockton_GRAD (0.0158)	ThePathYearbook (456)
SET_Stockton (58)	StkCollAlumni (0.0157)	stocktonospreys (375)
stocktonospreys (56)	stocoargo (0.0154)	RSCGrantsOffice (361)

Table 3. CSUSM network accounts ranked in order of decreasing value of the metrics. The metric value is included in parentheses next to the account name. The median values are in parentheses next to their corresponding metric in the column title.

In-degree (1)	Eigenvector centrality (0.0027)	Betweenness centrality (0.8)
csusmnews (174)	csusmnews (0.0232)	csusmnews (9691)
CSUSMCC (119)	CSUSMCC (0.0191)	CSUSMCC (3237)
csusmHOPE (109)	csusmHOPE (0.0187)	csusmHOPE (2601)
CSUSM_USU (91)	CSUSMEL (0.0166)	CSUSM_USU (1351)
CSUSMEL (85)	CSUSM_USU (0.0163)	CSUSMEL (1250)
ASI_CSUSM (75)	ASI_CSUSM (0.0157)	CSUSMAdvising (858)
CSUSM_Greeks (67)	CSUSMAdvising (0.0157)	CSUSM_Greeks (604)
CSUSMAdvising (63)	CSUSM_Greeks (0.0147)	ASI_CSUSM (603)
CSUSMdiningserv (57)	csusm_gradstudy (0.0140)	csusm_chabss (479)
csusm_chabss (51)	CSUSMdiningserv (0.0131)	CSUSMdiningserv (465)

Table 4 identifies the accounts whose values rank them in the top ten for each metric category. There are five in the RSC network and nine in the CSUSM network. Each primary cluster found in the networks contains a least one of these important accounts or the library account itself. That is, each cluster plays host to at least one important hub, which wields considerable influence over the cluster.

Table 4. Accounts ranked in the top ten for all three metrics

RSC account name	Account description	CSUSM account name	Account description
Stockton_edu	Primary institution	csusmnews	Primary institution
RSCCampusCenter	Campus center	CSUSMCC	Career center
stocoargo	Student newspaper	csusmHOPE	Student health center
Stockton_GRAD	Primary graduate programs	CSUSM_USU	Student union
StkCollAlumni	Alumni	CSUSMEL	Academic community outreach
		ASI_CSUSM	Student organization
		CSUSM_Greeks	Fraternity & sorority orgs.
		CSUSMAdvising	Undergrad. advising center
		CSUSMdiningserv	Campus dining services

DISCUSSION

Each of the three metrics used in this study indicate that institutional accounts, such as those maintained by the college/university, departments, and programs, are the most influential accounts in both the RSC and CSUSM follower/followee Twitter networks. The fact that this is true for the networks of both libraries suggests that it is a general feature of academic library Twitter networks, at least for medium sized, primarily undergraduate institutions like RSC and CSUSM. Further, these accounts are also likely to be influential in the wider institutional network, making them vehicles for information dissemination beyond the direct reaches of the library account. Recall that the metrics measure different definitions of importance. In-degree is the total number of incoming connections to an account; it represents the amount of attention the account receives. Eigenvector centrality is based on the premise that connecting to influential accounts bestows more influence than connecting to unimportant accounts. Finally, accounts with high betweenness centrality act as bridges; they exist on many short paths between vertices, which can allow information to reach many with just a few retweets. Accounts with high values of each of these metrics are connected to many highly connected accounts and can allow information to pass efficiently through the network. This study demonstrates that institutional accounts in the RSC and CSUSM library networks have these properties.

Interestingly, the most influential accounts are not those of faculty, staff or students. Influential accounts, generally, are also not the ones with the largest number of total followers, nor are they the most active on Twitter. The accounts with the most total number of followers are generally those that are known nationally or internationally, *e.g.*the Library of Congress and Hootsuite (a social media management platform), but are not necessarily central to the library networks. The most active Twitter accounts are those of individuals. There are two exceptions to this in the CSUSM network. The official account of the institution (@CSUSMnews) ranks highly in total number of followers, and @CSUMEL is the most active on Twitter (based on total number of tweets).

The metrics used in this study determine influential accounts based on the structure of the network, *i.e.* based on the wiring of the connections. Institutional accounts are wired to be important. Kim et al. (2012) discovered that the institutional accounts retweeted more library information than any other group; their behavior is significant. Combining these findings, one can conclude that such accounts are of paramount importance to the library networks. Their placement in the network sets them up to be influential (current findings). Then, their behavior allows them to capitalize on their position (Kim et al.). The results from the present study also highlight the importance of active participation by a library's account to the institutional Twitter conversation. As an institutional account itself, it is in a prime position to disseminate information from sister accounts within the college or university. This is also supported by the findings of Kim et al. Their study determined that, in addition to retweeting the most library information, the institutional accounts spawned more retweets (of their retweets) than any other group.

This information can be exploited by librarians for information dissemination purposes.Yep and Shulman (2014) show that partnering with highly connected accounts and agreeing to retweet messages can significantly increase exposure to library content. For example, if the library is hosting an event of interest to the campus as a whole, it might contact the primary account for the institution and request that its messages get retweeted. Both the primary accounts for RSC and CSUSM have many followers, most of which are affiliated with the institution and are potential recipients of the information. Alternatively, if the library obtains access to a new literature database, it could contact the account for the literature department. A simple click of the retweet button by the literature account will pass the message directly to those who are interested. Of course, it is not suggested that this be the only method of outreach. The library would probably do well to contact the literature department and faculty directly; however, using Twitter in this way requires minimal investment and will likely result in a more direct line to many literature students.

These examples, and the densely connected networks shown in Fig. 3 and Fig. 4, suggest that the reach of a library's Twitter program extends beyond its direct followers. Therefore, the number of followers is a poor estimation of the value or impact of a library's Twitter account. The environment in which it resides, *i.e.* the network, can play a significant role. Messages are capable of spreading broadly if retweeted by influential accounts, which are well connected in the library network and have many connections outside of it. The methodology presented above identifies these accounts and, notably, enables the quantification of the rather abstract concept of importance.

CONCLUSIONS

This work describes an analysis of the follower/followee Twitter network for academic libraries housed within two medium sized, primarily undergraduate institutions. Three network metrics, describing different measures of importance, were calculated for each account in the networks. This allowed for the influence of each account to be quantified. By all measures, institutional Twitter accounts were found to be the most influential in both networks, implying that this is a general feature of Twitter networks of academic libraries. The library Twitter network is a subset of a broader network encompassing the institution and its other divisions. The smaller library network was used to determine the significance of the institutional accounts; however, their influence extends beyond the confines of the library's Twitter neighborhood. Thus, the institutional accounts can be used to propagate information into the broader network of the institution. Partnering with the appropriate accounts can quickly direct information to a target audience.

This study shows that influence commanded by the institutional accounts is granted by the wiring of the connections in Twitter. Kim et al. (2012) found that such accounts retweet more library information than any other group. Thus, their behavior allows them to capitalize on their position in the network. Well connected, yet silent, accounts cannot propagate information, nor can vocal but poorly connected accounts. The results of Kim et al. combined with those of the present study demonstrate that institutional accounts are both well connected and vocal. The library, as an institutional account itself, can play an important role in information dissemination within the institutional Twitter network. This work, like the others that came before it, suggests that librarians managing Twitter profiles interact with other accounts on campus and take advantage of Twitter-specific features in order to contribute to the institutional Twitter conversation.

Librarians can adopt the simple procedure outlined above to map the networks of their home libraries, identify influential accounts, likely institutional, and rank them. Dickson and Holley (2010) note the time intensive nature of social media activities. It's true that managing a successful social media campaign requires a considerable investment in time; however the analysis described here does not contribute to it. Apart from the data download, which can take a few days while running in the background, the automated analysis with NodeXL can be completed in a matter of seconds with only a few clicks of the mouse. Such an analysis can allow librarians to tailor their Twitter outreach activities. In particular, it can complement a detailed examination of followers such as the one completed by Sewell (2013). One of the drawbacks of a microscopic examination of followers is the time requirements for manually obtaining follower information. NodeXL can help automate the import of much of the information, *e.g.* the numbers of followers and tweets, date of account creation, account description, among others.

ACKNOWLEDGEMENTS

This work is dedicated to the memory of Barry Alexander Moores. The authors would like to thank Talitha Matlin and Carmen Mitchell from the Kellogg Library at CSUSM for assistance interpreting data from their institution. This work benefitted from discussions with Mary Ann Trail and Stephen Tsui. The authors also appreciate the Social Media Research Foundation for producing and providing access to NodeXL, the software package used in the study.

REFERENCES

1. Cassidy, E., Britsch, J., Griffin, G., Manolovitz, T., Shen, L., & Turney, L. (2011). Higher education and emerging technologies: Student usage, preferences, and lessons for library services. Reference & User Services Quarterly, 50(4), 380–391.

2. Cassidy, E.D., Colmenares, A., Jones,G.,Manolovitz, T., Shen, L., &Vieira, S. (2014).Higher education and emerging technologies: Shifting trends in student usage. The Journal of Academic Librarianship, 40(2), 124–133.

3. Cole, S. (2009). 20 ways for librarians to use Twitter. Library Journal, 134(11), 25.

4. Cuddy, C., Graham, J., & Morton-Owens, E.G. (2010). Implementing Twitter in a health sciences library. Medical Reference Services Quarterly, 29(4), 320–330.

5. Del Bosque, D., Leif, S.A., & Skarl, S. (2012). Libraries atwitter: Trends in academic library tweeting. Reference Services Review, 40(2), 199–213.

6. Dickson, A., & Holley, R.P. (2010). Social networking in academic libraries: The possibilities and the concerns. New Library World, 111(11/12), 468–479

7. Easley, D., & Kleinberg, J. (2010). Networks, crowds, and markets: Reasoning about a highly connected world. New York: Cambridge University Press.

8. Glazer, H. (2009). Clever outreach or costly diversion? An academic library evaluates its Facebook experience. College & Research Libraries News, 70(1), 11–19.

9. Gunton, L., & Davis, K. (2012). Beyond broadcasting: Customer service, community andinformation experience in the Twittersphere. Reference Services Review, 40(2),224–227.

10. Hansen, D., Shneiderman, B., & Smith, M.A. (2010). Analyzing social media networks with NodeXL: Insights from a connected world (1st ed.). Burlington, MA: Morgan Kaufmann.

11. Kim, H.M., Abels, E.G., & Yang, C.C. (2012). Who disseminates academic library information on Twitter? Proceedings of the American Society for Information Science andTechnology, 49(1), 1–4.

12. Milstein, S. (2009). Twitter for libraries (and librarians). Computers in Libraries, 29(5), 17–18.

13. Newman, M. (2010). Networks: An introduction (1st ed.). Oxford; New York: Oxford University Press.

14. Park, J. -H. (2010). Differences among university students and faculties in social networking site perception and use: Implications for academic library services. The Electronic Library, 28(3), 417–431.

15. Petrovic, S., Osborne,M.,& Lavrenko, V. (2011). RT towin! Predictingmessage propagation in Twitter. Fifth International AAAI Conference on Weblogs and Social Media (Retrieved from

16. Sewell, R.R. (2013).Who is following us? Datamining a library's Twitter followers. Library Hi Tech, 31(1), 160–170.

17. Suh, B., Hong, L., Pirolli, P., & Chi, E.H. (2010).Want to be retweeted? Large scale analytics on factors impacting retweet in Twitter network. 2010 IEEE Second International Conference on Social Computing (SocialCom) (pp. 177–184).

18. Yang, J., & Counts, S. (2010). Predicting the speed, scale, and range of information diffusion in Twitter. Fourth International AAAI Conference on Weblogs and Social Media

19. Yep, J., & Shulman, J. (2014). Analyzing the library's Twitter network: Using NodeXL to visualize impact. College & Research Libraries News, 75(4), 177–186.

6
CHAPTER

Knotworking in Academic Libraries: Two Case Studies from the University of Helsinki

Yrjö Engeström, Heli Kaatrakoski, Pälvi Kaiponen, Johanna Lahikainen, Anne Laitinen, Heli Myllys, Juhana Rantavuori, Kaisa Sinikara

University of Helsinki, Finland

ABSTRACT

Librarians in academic libraries are facing major changes in their work due to, e.g., the internet, digitization, and increasing use of new channels for information retrieval by their most important clients, namely researchers. This creates challenges for librarians: both to deepen their own expertise and to develop innovative service models for their clients. In this paper we present a development project entitled 'Knotworking in the Library' from the Helsinki University Library. The project made use of the Change Laboratory method, which is an intensive developmental effort which facilitates improvements in the activities of organizations and changes in the organizational culture. The process started in Viikki Campus Library in 2009–2010 and continued in the City Centre Campus Library in 2010–2011. The aim was to create new kinds of partnership between libraries and research groups in the form of knotworking. By knotworking we mean a boundary-crossing, collective problem-solving way of organizing work. The knotworking model presented in this paper generated practical tools to assist selected research groups in dealing with data management related-issues.

Keywords: Knotworking, Change Laboratory, service innovation, research data management

INTRODUCTION

Librarians in academic libraries are currently facing major changes in their work due to, e.g., the internet and digitization of material. Also, research funders increasingly demand a detailed description of the research data created and a plan for their lifecycle management. Researchers' resources may not be sufficient to meet these new requirements. The need for new kinds of support services has motivated academic libraries to search for models whereby librarians and research groups develop new ways of collaborating to meet the challenges (Brophy, 2007; Earnshaw and Vince, 2007; Rader, 2002). The Helsinki University Library has taken a proactive approach to developing services for research groups. The new tasks include organizing, archiving and storing huge amounts of research data as well as viewing the publishing processes in a new way. In this paper, we present a two-phase change effort aimed at addressing these challenges. In both change efforts — one in Viikki Campus Library in 2009 and the other in the City Centre Campus Library in 2010 — innovative services were developed through close collaboration between the librarians and the selected research group members.

Knotworking and the Change Laboratory

The Helsinki University Library has started addressing the challenges in a concrete way through knotworking (Engeström et al., 1999). Knotworking is a boundary-crossing, collective way of organizing work. Knotworking is a means for participants to create continuity in the production of the shared object (e.g., research data archiving). According to Engeström the knot symbolizes a rapidly pulsating, distributed and partially improvised collaboration between loosely connected actors and activity systems (Engeström, 2000, p. 972). Knotworking is "a longitudinal process in which knots are formed, dissolved, and re-formed as the object is co-configured time and time again, typically with no clear deadline or fixed end point" (Engeström, 2000, p. 973). In other words, knotworking differs from team work in the sense that continuity is connected to the object, not the practitioners; the practitioners and the initiators of knots can change.

Applying the knotworking model requires a long-term effort. This can be achieved by using the Change Laboratory method based on the methodology of Developmental Work Research (Engeström, 2007). The Change Laboratory method can be used to effect changes in organizational culture as well as to continuously improve the activity of organizations. This requires the active participation of the practitioners in an intensive process: normally 5–10 sessions are arranged weekly or fortnightly. All the sessions are video and audio recorded, and the recordings are used for reflection in later sessions and in research.

The Change Laboratory process begins by collecting ethnographic data such as video recordings and interviews (so-called mirror material) representing the present state of the work practices. In the sessions the participants analyze the challenges in their work practices with the help of the mirror with the aim of revealing historically developed, work-related tensions and disturbances and of planning a new model of activity to solve them. Various analytical models and tools such as the cyclical model of expansive learning (Engeström, 1987, p. 323) and the triangular representation of an activity system (Engeström, 1987, p. 78) can be used to help practitioners perceive their work activity as a whole. In the Change Laboratory sessions practitioners move between the past, present, and future states of their activity, which brings a historical element into the creation of something new. The Change

Laboratory produces the rich data about collective problem solving and collective learning that change management requires. The method has been used for over ten years to help organizations such as hospitals, schools, newspapers, theatres, factories etc.

The Helsinki University Library

Helsinki University Library is Finland's largest multidisciplinary university library and it provides information and library services across a number of disciplines at four university campuses. The Helsinki University Library aspires to be one of the leading multidisciplinary research university libraries in the world by 2020.

Helsinki University Library in its present state was founded on the 1st of January 2010 after a long period of development lasting from 1993 to 2010. In 1993 there were 160 libraries. An international evaluation in 2000 and a follow-up evaluation in 2004 were crucial factors in creating the present model of a single university library. The new Helsinki University Library consists of the following centralized library services: 1) administration and development, 2) acquisition and metadata services, and 3) digital services.

In addition four subject-specific campus libraries (in Meilahti, Kumpula, Viikki and City Centre) serve teaching and research. These came into being out of the earlier faculty and department libraries during 1999–2009. Each of them had a long history of its own. The structural changes presently underway in the City Centre Campus Library will be completed in 2012 when the new Kaisa Building will open its doors, bringing together most of the smaller libraries which are currently scattered over eleven locations.

Since 1st August 2006 the former Helsinki University Library is the National Library of Finland. Like the former University Library, it is an independent institution, which is of national relevance within the university as a national "memory bank" and which provides services to the entire national library network. The university library duties are laid down in the Service Agreement 2010–2012.

Research Site

Two different campus libraries were selected for this research project. The Viikki Campus Library was selected as the first pilot. This library was the oldest campus library (1999–) and a tradition of collaboration with research groups had already been established. The faculties on the Viikki campus are Agriculture and Forestry, Biological and Environmental Sciences, Pharmacy and Veterinary Medicine. Viikki is a unified campus with many active research groups of international repute.

The City Centre Campus Library was the opposite of Viikki. As it was established only in 2010, merging five faculty libraries and the undergraduate library, it was much younger, and it is also a very different type of campus. The faculties at the City Centre Campus include Arts and Humanities, Behavioural Sciences, Law, Social Sciences and Theology. The humanities and social sciences do not seem have a tradition of working in research groups as the biosciences campuses do. The resources for research are very different too. Humanities and social sciences are book-based whereas the biosciences are heavy users of net-based material. In fact, the concept of a “research group” remained elusive to most of the researchers at the City Centre Campus. Another important reason for selecting the City Centre Campus Library was that the library was under construction both inside and out. The new building will not be ready until 2012, which means that there was some time to develop the services.

Features of the Viikki Campus Library

The Viikki Campus Library was founded in 1999 when four faculty libraries and smaller units were merged and moved from several addresses in the city to a new library building in Viikki. Therefore, the library has had time to adjust to the changes brought about by the merger. The disciplines based at the Viikki campus traditionally conduct their research in research groups. The Library had already taken an active role in serving the researchers and had been taking part in development groups for new requirements for research material. The librarians were subject experts and had according responsibilities. They were well connected both inside the university and outside. Several research institutes had also moved to the campus to further research-related cooperation with faculties and institutions.

At the time of the Change Laboratory in 2009, Viikki employed 39 permanent staff members; it boasted 12,500 shelf metres of books and monographs, 3,100 periodicals, 18,000 e-journals, 250 databases, approximately 2,000 electronic books, 200 study places and 100 workstations for the clients. The library floor space is 5,000 square metres on four floors.

Librarians and Research Groups in the Viikki Campus Change Laboratory

Fifteen professionals from the Viikki Campus Library, including the library director, participated in the Change Laboratory sessions. It was the intention of the project to involve professionals from different fields of expertise in order to reflect the multi-voicedness of the community.

A set of mirror data was collected prior to the sessions. Mirror data are video clips from interviews with librarians and the members of the participating research groups. The video clips depict in a concrete way possible disturbances and challenges in the work activity of librarians. The mirror data from Viikki consisted of interviews with two librarians and one representative of each research group. The interviews aimed at unearthing changes in library work and its object, and the possible role libraries could play in supporting research groups in the future. One of the two librarians interviewed had worked for the library for over ten years; the other had a particular interest in the potential of ICT in further developing library services. All of the interviews were videotaped and the researchers selected key clips to be shown as mirror data in the Change Laboratory sessions.

The library staff participating in the Change Laboratory sessions chose two different research groups in terms of research interest and group structure for the pilot project. These were Peatland Ecology (Department of Forest Sciences) and the Cyanobacteria Group (Food and Environmental Sciences).

In the course of the Change Laboratory sessions it became clear that both groups had a lot in common with regard to their scientific scope. Both research groups were represented in the sessions by five to ten researchers who were in different stages of their academic careers. The leaders of both research groups were present in the sessions. In the interest of obtaining mirror data, two senior researchers from the Peatland Ecology Group and the leader of the Cyanobacteria Group were interviewed. The Change Laboratory consisted of six two-hour sessions in September–November 2009.

Features of the City Centre Campus Library

After the pilot in the Viikki Campus Library with two research groups, the second phase of the 'Knotworking in the Library' project was conducted in the City Centre Campus Library with four research groups. The City Centre Campus Library became operational at the beginning of 2010 and was in the middle of its transformation from a number of smaller libraries to one central facility.

The City Centre Campus Library serves five faculties in the city centre: the Faculty of Arts and Humanities, the Faculty of Behavioural Sciences, the Faculty of Law, the Faculty of Social Sciences, and the Faculty of Theology. At the time of the project, the library was still scattered over 11 locations. Because of the disciplines involved and, perhaps, the geographic dispersion, it seems that the researchers at City Centre

Campus do not identify themselves as members of research groups but see themselves as individual researchers or, at most, identify with the department involved. The new Kaisa library building will cover an area of 13,000 square metres and have about 1.5 million books. The library expects to receive about 1.1 million visitors annually.

Librarians and Research Groups in the Centre Campus Change Laboratory

Eighteen librarians from the City Centre Campus Library representing different occupations participated in the Change Laboratory sessions and four of them were interviewed beforehand. Four research groups from the City Centre Campus Library participated in the project. Five groups were interviewed, because one group from the Finnish Language Department withdrew from the project after the interviews.

The research groups were: Cognitive Science from the Faculty of Behavioural Sciences; The Construing Reader in the Framework of Media Concept Project (Finnish Language) from the Faculty of Arts, New Media Environment; Challenges for Copyright from the Communication Law in the Faculty of Law; and Politics of Philosophy and Gender (Gender Studies) from the Faculty of Arts. One to four researchers from each research group participated in every session and from three to five of them were interviewed beforehand. The Change Laboratory consisted of eight weekly sessions in October and November 2010.

Knotworking in the Library

Knotworking can be understood in terms of an innovative model of organizing work in which professionals and their clients form 'knots' in order to work on a shared object. These knots are less fixed than teams and can adapt to changes more rapidly. New knots can be formed when circumstances so require. In the context of the present project knotworking was initiated in the Change Laboratory sessions. In both cases the knotworking model led to continuation of the work independent of the Change Laboratory sessions, and led to the creation of a service tray as the central outcome of both of the phases of the project. The two case studies in this paper represent different types of knotworking.

Viikki Campus

In this article we take a closer look at the creation of a data management plan (DMP) in a joint effort by a researcher of the Cyanobacteria Group and the library staff. This service innovation was prompted by the researcher's imminent need to submit such a plan in the context of a grant application to the Academy of Finland, which had recently introduced a DMP requirement. Figure 1 depicts the timeline of the creation of the data management plan.

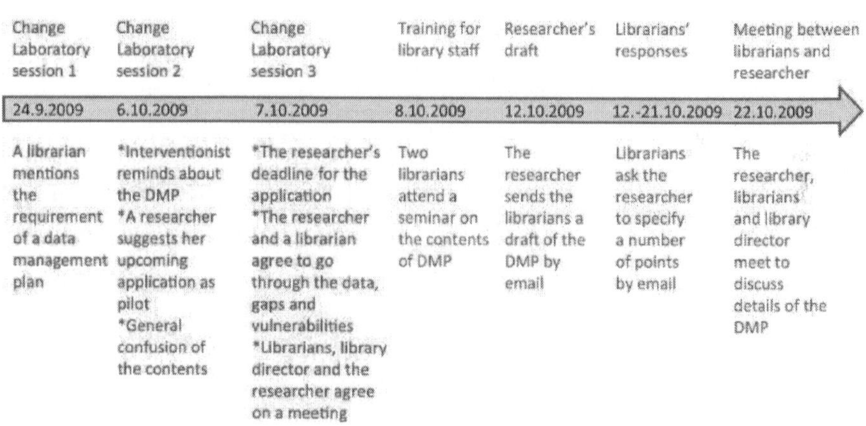

Figure 1. Timeline of the co-configuration of a pilot data management plan by the Viikki Campus Library staff and a researcher of the Cyanobacteria Group

The Change Laboratory process in Viikki consisted of six sessions. The first session was only attended by the library staff. When discussing possible new areas of collaboration between the library and the selected research groups, one librarian mentioned the new requirement set by the Academy of Finland which stipulates that research groups and projects must draw up a data management plan. The librarian suggested that the library use its expertise to act as an intermediary between the research groups, the Academy of Finland and different administrative parties involved to meet this requirement, even though it was as yet unclear what this would entail.

The library staff and representatives from both pilot research groups participated in the second Change Laboratory session. When the interventionist reminded the group of the data management plan requirement, a researcher from the Cyanobacteria Group commented that the present procedure was unclear: 'I have done something like this [in the past], but I'm not entirely sure what it actually means. I've usually just written a paragraph…' The researcher then mentioned that she was, in fact, about to apply for a grant from the Academy of Finland: 'This reminds me: the deadline for the next applications is at the end of the month, so is it OK for me to let them know now that I'm working with the library to produce such plan?' The researcher had an actual need for a new tool to meet the application requirements and she seized the opportunity to collaborate with the library. The procedure was new for the library staff too: 'As I've understood it, the Academy has not given any specific instructions as of yet. We're off to a seminar actually [with a colleague] the day after tomorrow. The purpose is to hear about the data management plan.'

The idea was further elaborated in the third Change Laboratory session which was devoted to the particular needs of the Cyanobacteria Group. In response to the interventionist's reminder about the data management plan, the Cyanobacteria researcher who had in the previous session briefly mentioned her imminent application to the

Academy of Finland now emphasised the urgency of her situation: 'This is a very important issue. Personally, I need to have a plan as early as the 31st of October, so my share of the pilot case have to be completed before the end of the month.' The researcher also acknowledged the need to design the data management plan in such a way that it could later be expanded to serve the whole research community. The library director responded positively: 'Great! Why don't we set a date for our best experts and your best experts to work on this?' The researcher and three librarians then took concrete steps towards the co-configuration the data management plan by setting a date for a meeting to go through the details.

The data management plan was a learning experience for all parties involved: the library staff attended a seminar after this initial step in the budding collaboration. The researcher sent a draft of the data management plan to the librarians and an exhaustive exchange of emails followed: the librarians were negotiating amongst themselves the best way to proceed. To get an overall picture they sent the researcher a list of questions in an attempt to clarify the specific nature of the data, the storage space and software needed, usage rights and data protection issues.

After the researcher's responses had been received, a meeting was organized between the researcher and a colleague from the same group, eight librarians and the library director. For the researcher one of the benefits of collaborating with the library staff was that it helped her organize her own data, detect possible overlaps and improve the description of the data. To illustrate this, Figure 2 contains a flow chart created by the researcher. This flow chart shows what types of data are involved and where the data, statistical and metadata files are deposited.

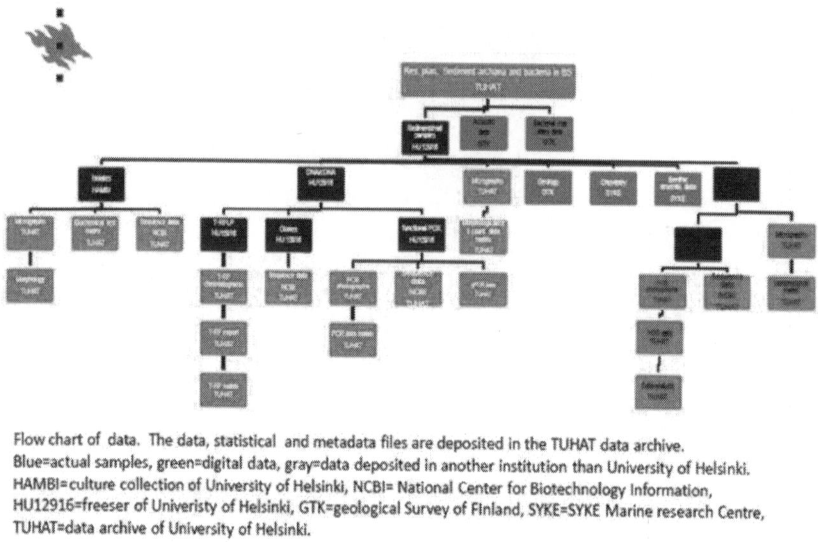

Figure 2. Flow chart extracted from a data management plan required by the Academy of Finland and created by a researcher from the Cyanobacteria group.

It was our intention to describe a second test case, a data management plan for a Peatland researcher, but due to the researcher's timetable this could not be done. Instead, the library professionals built a website for the Viikki campus faculties with information about available data storage options, the data management requirements of the Academy of Finland, and other important issues. This data management website has been updated and expanded ever since by the library staff and now also includes information for the **City Centre campus.**

City Centre Campus Library

The second knotworking case we present in this paper produced a quick reference guidebook for the Cognitive Science research group. This phase of the project was markedly different from the Viikki phase:

knotworking was sparked by and initiated in the Change Laboratory sessions, but it mainly took place outside the Laboratory.

At the beginning of the Change Laboratory process it became evident that the library was unable to meet the data management needs of the Cognitive Science research group. Thus, innovative solutions were needed.

Some of the research group's concerns involved general threats to stored data. A research expressed it as follows: 'Yes, today we had a fire drill. This is an example of a possible threat, because our data would have burned. I mean, everybody grabbed their own laptops but did anyone think of taking the external hard drive along?'

Other concerns focused on data management issues at a local level. The following excerpt highlights some of the problems associated with a lack of shared data storage practices.

'Researcher 1: 'Yes, we store them [data] on several different computers.

Librarian 1: On a number of different computers, and each researcher has his own index system and there is no common format (...)

Researcher 2: We store everything on external hard drives.

Librarian 1: Yes.

Researcher 2: So if you want one particular file, then you ask a colleague on which drive it is.

Librarian 1: Yes.

Researcher 2: And the answer is that it's probably on that drive.

Librarian 1: Yes.

Researcher 2: And then you look, it could be here, it could be there. Before I left for my holidays I made back-ups of everything I'm involved in and it took the whole day to track the files down, I mean, finding out which computer was used and how the data are registered.'

This excerpt reveals that storage, encoding and description practices were rather random, varied from one researcher to another, and, as the head of the department stated 'may lead to not being able to use perfectly valid data due to serious shortcomings in storage practices'.

The library professionals and Cognitive Science researchers began knotworking to develop solutions for the data management problems; the timeline of the creation of a quick reference guide is shown in Figure 3.

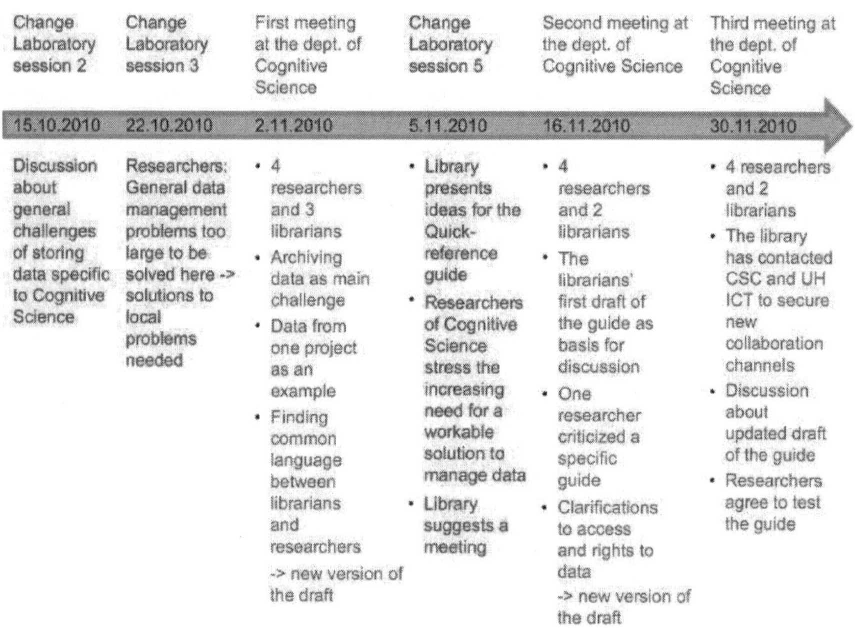

Figure 3. Timeline of the creation of a quick reference guide by the City Centre Campus Library staff and the researchers of the Cognitive Science research group.

The issues of the Cognitive Science researchers were discussed in more detail in the second Change Laboratory session: it became apparent that they were having serious problems managing, storing and archiving their data. The research group handles mostly large video files, such as EEG files and eye movement recordings, and these are difficult to store, that is, the available options were deemed insufficient. During the second Change Laboratory session the library suggested that the research group make use of the Finnish Social Science Data Archive (FSD). The researchers were not convinced right away, as was demonstrated by the third Change Laboratory session, when the discussion centred on the researchers' need to sort out their own, local practices first before zooming in on large-scale data management issues. The librarians and researchers decided to organize a special meeting to enable the librarians to familiarize themselves with the work methods of the researchers.

The first of such meetings in a series of three was held in the department of Cognitive Science and was attended by three library professionals and four researchers. First, the participants had to find a common language: the researchers had brought some of their data to illustrate the particular issues involved in archiving, tagging and describing the content — preferably in such a way that the data would be understandable independently of the researcher.

The fourth session was devoted to the needs of another research group. Then, in the fifth Change Laboratory session, the library staff presented their ideas about what they now called a quick reference guide. This term was used to denote a collection of guidelines to be used at different stages of the research to store data systematically and to do away with random data descriptions used by tens of individual researchers. The library explained that the purpose of the guide was to enable data to be stored safely in a structured way. The researchers welcomed this type of assistance and a meeting was organized outside of the Change Laboratory to discuss the details of the quick reference guide.

The second meeting outside the Change Laboratory context took place at the department of Cognitive Science. This time two professionals from the library and four researchers from the department participated, one of which (Researcher 2 in the excerpt below) had not been present at the Change Laboratory session or the previous meeting with the librarians. On the table was the latest draft which had been emailed to the researchers before the meeting. One researcher had reservations about the need for the guide and expressed his fear that the guidelines might turn out to be too rigid:

'The bad thing about instructions is that they are hardly ever updated and soon lose touch with actual practices. Methods are developed and changed all the time, new things are invented and new stuff is being tried out so if it [the guide] is too detailed it will be outdated very soon.' (Researcher 2)

After hearing the librarians' comments, Researcher 2 offered some advice for improving the guide:

'The most important thing about such a guide is that the issues [a researcher] easily overlooks are dealt with. These are obvious things, such as who has collected the data and when; who has the right to use the data, who has the right to publish them in the future and so on.'

The ensuing discussion between the researchers and the librarians showed that many of the current data practices were not documented and were implicit. The information provided by the researchers enabled the librarians to prepare a new draft which emailed to the researchers and then discussed at the third meeting a fortnight after the second one. The meeting was attended by the same two librarians and four researchers as the second meeting. They went through the guide in detail and one of the researchers agreed to test the usability of the guide and report back to the library.

In order to sort out issues of large-scale storage of data, the library had been in touch with the IT Center for Science (CSC). CSC is a Finnish non-profit organization which provides ICT support to research institutes, e.g., network, application and data management services. The librarians introduced the option of storing the data at CSC at the meeting with the researchers, but this could not be decided upon, partly because some junior researchers required approval from the head of their research group. In the end, it was decided that the librarians would stay in touch with the researchers and help them fill in the appropriate application form.

As a result of knotworking between the library professionals and the researchers of Cognitive Science a quick reference guide was in fact produced. Upon the recommendation of the librarians, the guide is structured in three sections, each covering a specific aspect of data management during different stages of a research project. The outline of the quick reference guide is as follows:

1. Section 1. Description of the study and the data
    1. Section 1.1. Meta documents for the study and the data
    2. Section 1.2. Possibilities for further development of metadata (documentation) of research process and research data
2. Section 2. Description of individual experiments
    1. Section 2.1. Meta documents of the experiment
    2. Section 2.2. Possibilities for further development of individual experiment data
3. Section 3. Description of recordings
    1. Section 3.1. Meta documents of the recordings

2. Section 3.2. Possibilities for further development of recordings metadata

The final draft of the quick reference guide is a rather extensive set of guidelines which needed to be tested in practice by the researchers. This was done in early 2011. The guide was recommended by a cognitive sciences researcher also to be introduced to the neurosciences researchers' use.

The collaboration between the library staff and the researchers in creating the quick reference guide is an illustrative example of knotworking in solving data management problems. The problems were identified by the researchers and the librarians in the Change Laboratory sessions but the creation of the quick reference guide mainly took place in meetings organized by the practitioners. The participating librarians or the researchers were not always the same people in the Change Laboratory sessions or in the separate meetings: the collaboration was built on the shared object and the practitioners involved varied.

At a wider level the knotworking model created during the project included the innovation of new way of working. Instead of serving clients in the library, librarians went out of the library to meet researcher in their departments and offices. Furthermore, librarians cooperated actively with researchers in order to develop services researchers needed. This was a new way of working not only for librarians but also for researchers.

CONCLUSION

The two phases of the 'Knotworking in the Library' project differed in terms of their starting points. Viikki is a Biosciences campus with a long tradition of working in research groups and little need for printed resources. It is also worth noting that the leaders of the research groups were present at all the Change Laboratory sessions.

The disciplines represented in the City Centre Change Laboratory, particularly humanities and social sciences, depend more on printed texts and monographs. These disciplines traditionally do not work in research groups but the researchers act as individuals. The participants in the four pilot groups were at different stages of their academic careers, but the most senior researchers and group leaders were typically not able to attend the Laboratory sessions due to practical circumstances. This led to a situation in which junior researchers did not have the authority to make decisions.

Regardless of these different starting points, both processes led to closer collaboration between researchers and the library staff.

One challenge to the library today is that researchers may either not recognize the relevance of library services or may not even know about them. In this particular project the library took on the challenge from both angles: in Viikki the library staff took up the challenge of creating a data management plan. Meeting the researchers' needs involved a lot of communication between the researchers and the library, amongst the library staff themselves and between the library and other partners. It was the researchers that took the initiative in a Change Laboratory session, but the meeting to sort out the details was initiated and hosted by the library. In the case of City Centre Campus Library the librarians went outside the confines of the library and reached out to the researchers with their knowhow. They met with the researchers at their place of work and closely worked with their colleagues in between the meetings.

In both cases the innovative services came about through knotworking. In Viikki the object of knotworking was a data management plan. Researchers and library professionals worked together to build a new tool which could be tested and then expanded and developed to serve a larger body of users. The individual library staff differed at the various stages of the process. Likewise, in City Centre Library, the knot was formed to work on the quick reference guide. There were always 2–3 library professionals and 4 cognitive scientists involved in creating the quick reference guide.

The two phases of the 'Knotworking in the Library' project proved to be a demanding process. The two different research sites differed in terms of context and thus generated different innovations. In Viikki the process generated a comprehensive service tray which included the data management plan described here. The service tray also included 1) training and briefing related to the new information system, 2) tracking new releases and training the use of new tools for this purpose, 3) offering news flashes of new services and tools at the departments, and 4) assistance in issues related to visibility of the research group.

In the City Centre Campus Library the earlier experiences from Viikki could be put to good use: here too a tool for tracking new releases was implemented in addition to the quick reference guide. The City Centre Campus Library is currently changing its organizational structure to support the knotworking model: librarians must find their own way of collaborating with the research groups on the campus.

Envisioning the Future: Different Types of Agency within the Two Knotworking Cases

The new model of knotworking in academic libraries requires a new kind of agency from all parties involved. The two Change Laboratory processes presented here may not have led to large-scale breakthroughs, but they certainly illustrate how demanding it is to change one's own working method or organization. The two cases demonstrate that without the observations that led to the creation of these innovative solutions it might not have been possible for the librarians to critically evaluate their working method. Without these observations the pressure to change might not have been felt.

In the future it will be interesting to investigate more closely how initial ideas evolve into practical solutions and what kinds of agentive/initiative this requires. Engeström and Sannino (forthcoming) maintain that 'the formation of an expanded object and corresponding new pattern of activity requires and brings about collective and distributed agency, questioning and breaking away from the constraints of the existing activity'. This requires new forms of agency so that it becomes possible to move 'from momentary individual initiative to sustained collective effort' (Engeström & Sannino, forthcoming).

This could be analysed through the six main types of participants' emerging agency that Engeström has traced within earlier Change Laboratory interventions (Engeström, 2011). These types can be briefly characterized as follows:

1. Criticizing the current activity and organization.

2. Questioning or resisting the interventionist or the management. Resistance towards the intervention process or the management may take the form of criticism, questioning, opposition and rejection.

3. Explicating new possibilities or potential in the activity. This refers to reflecting on previous positive experiences and seeing them as a potential left unacknowledged or characterizing the problematic object as a source of new possibilities.

4. Envisioning new patterns or models of the activity. Envisioning can occur in the form of preliminary suggestions or presentation of comprehensive models for the future.

5. Committing to concrete actions aimed at changing the activity and expressing this through commissive speech acts.

6. Taking consequential actions to change the activity. Participants of the interventions may actually change the

situation through a sequence of actions. This may occur in, between and after the laboratory sessions.

REFERENCES

1. Brophy, P. (2007): 'Communicating the library: Librarians and faculty in dialogue',Library Management, 28, 515–523

2. Earnshaw, R.E. and J.A. Vince, eds. (2007): Digital Convergence: Libraries of the Future. London: Springer.

3. Engeström, Y. (1987): Learning by Expanding. An Activity Theoretical Approach to Developmental Research. Helsinki: Orienta-Konsultit Oy.

4. Engeström, Y. (2000): 'Activity theory as a framework for analyzing and redesigning work', Ergonomics, 43(7), 960–974.

5. Engeström, Y. (2007): 'Putting Vygotsky to work: The Change Laboratory as an application of double stimulation'. In: H. Daniels, M. Cole and J.V. Wertsch (eds.),The Cambridge companion to Vygotsky. Cambridge: Cambridge University Press.

6. Engeström, Y. (2011): 'From design experiments to formative interventions',Theory and Psychology, 21(5).

7. Engeström, Y. and A. Sannino (forthcoming): Volition and agency in organizations: An activity theoretical perspective.

8. Engeström, Y., R. Engeström and T. Vähäaho (1999): 'When the center does not hold: the importance of knotworking'. In: S. Chaiklin, M. Hedegaard, and U.J. Jensen

(eds.), Activity Theory and Social Practice, Aarhus: Aarhus University Press.

9. Rader, H. (2002): 'Managing academic and research libraries partnerships', Library Management, 23, 187–191

7

CHAPTER

The Challenges facing Professional Education and Practice of Academic Librarianship in the HE Sector in the UK

Emilce Sandra Rees

ABSTRACT

This article examines the challenges facing professional education and practice of academic librarianship in a changing Higher Education world. Librarians in the UK, like all the professionals in the academic sector, are faced with changes in organisational mission within the university, and the changing role of information within society. Inter alia, they have to be accountable inside and outside the university and are in competition for resource allocation and funding. Against this socio-political background, the following eight challenges have been isolated: 1. recognition of librarianship as a profession (and the accreditation of library and information studies); 2. ethical professional responsibilities in a changing organisational world, and how they can be weaved into the librarian's professional education; 3. the librarian in a changing (HE) world; 4. the "assault" on professionalism; 5. the user's centrality: fact or fiction?; 6. the librarian in an era of global challenge: is resistance futile?; 7. Potemkin villages: fabrication and performance in the academy, and 8. feminisation of the HE professions and patterns of employment. 9. Lastly, I'll examine future scenarios for librarians and the implications of present practices, which affect all sectors.

INTRODUCTION

This article examines the challenges facing professional education and practice of academic librarianship in a changing Higher Education world. Librarians in the UK, like all the professionals in the academic sector, are faced with changes in organisational mission within the university, and the changing role of information within society. Inter alia, they have to be accountable inside and outside the university and are in competition for resource allocation and funding. Against this socio-political background, the following eight challenges have been isolated:

1. recognition of librarianship as a profession (and the accreditation of library and information studies);
2. ethical professional responsibilities in a changing organisational world, and how they can be weaved into the librarian's professional education;
3. the librarian in a changing (HE) world;
4. the "assault" on professionalism;
5. the user's centrality: fact or fiction?;
6. the librarian in an era of global challenge: is resistance futile?;

7. Potemkin villages: fabrication and performance in the academy, and

8. feminisation of the HE professions and patterns of employment.

9. Lastly, I'll examine future scenarios for librarians and the implications of present practices, which affect all sectors.

Recognition of librarianship as a profession

In order to gain professional recognition by practitioners at large and by CILIP(Chartered Institute of Library and Information Professionals), an academic librarian is subject to the following career passages: firstly, a 'placement' on a graduate trainee scheme after a first degree but before library school, which might not be the 'reflective practicum'. Guile (2005) envisages for "the development of professional artistry" but, more often than not, this is an opportunity for academic libraries to acquire highly qualified trainees to perform routine tasks for a very low salary. After this year, the average library trainee takes either a postgraduate Diploma or Master Degree at library school, usually accredited by CILIP. It is then expected that after two years in a professional post the librarian should want to apply for Chartership status which, in theory, guarantees professional recognition and has been described by CILIP as a "seal of approval", a landmark which determines the transfer from the theoretical to the practical. This 'micro-management' of the profession, and the need for independent years of 'training' is, however, counterproductive to its ambition to be of equal standing to the major liberal professions.

Traditionally, nobody enters librarianship to maximise personal prestige and income, but out of a sense of vocation. CILIP, the representative body for librarians, is not made up of what Freidson describes as "an elite of unusually successful and distinguished practitioners" but of lesser mortals (Freidson, 2001, p. 142). However, 'a restriction of trade' via selection might be professionally welcome if the result is recognition and a rise in official status. For example, in most European countries, librarians have to pass selective national exams prior to library school, which guarantees the trainees a civil servant post for life.

Where is the librarian professionally located? Freidson includes the librarian in his classification of the professions needing a professional degree (Freidson, 1986, p.53). In Bourdieu's Homo Academicus, librarians are mentioned only once in his taxonomy of French intellectuals, under "Book Trade", in the company of publishers and booksellers (Bourdieu, 1984, p. 265). In its most elementary sense, 'professionalism' should be judged by the degree in which the members of an occupation can make a living while controlling their own work, with a degree of autonomy and of self-regulation (Freidson, 2001, p. 17).

A profession is not "mature" until it questions its role: recognition of professionalism should be achieved via true reflective practice, which should not be allowed to become a panacea or another professional buzzword. The concept of reflective practice was popularised by Donald Schon, whose work focused on the theory/practice problem of professional education in American Universities (Schon, 1987, Guile, 2005). Contra Schon, Barnett argues that Schon's conception of the professional is too individualistic, as professional life is necessarily social and intersubjective (Barnett, 1997). Guile is also critical of 'reflective practice', as it perpetuates the theory/practice divide (Guile, 2005). However, 'reflective practice' is repeated like a mantra by library managers (usually at training events) when at root level a culture of conformity is demanded and adhered to by employees, since, as Chomsky states, "if you show too much independence and question the code of your profession too often, you are likely to be weeded out of the system of privilege" (Chomsky, 2000, p. 24). The notion of 'reflective practice' might also subvert "dominant orders of discourse" (Morley, 2003, p. 115) while the quality 'experts' demand conformity.

Librarians must position their role within the academic community (the librarian as a scholar has still professional validity in Europe). If their role is interpreted as peripheral, this might lead to extinction: eight subject librarians were to be made redundant in a Welsh University. Information World Review (2005, March, p. 9) declared in their defence, that "the librarian is not a support role". As Morley wonders whether academics are being de-professionalised from scholars to knowledge workers (Morley, 2003, p. 91), we should in turn question whether knowledge workers have also sunk a notch down, to be considered mere office workers or highly trained clerical staff. In the post-modern economy, all jobs seem to have been reduced to a set of skills [1] - while at the same time segregation is encouraged. Neo-taylorian practices such as the "three e's" (economy, efficiency and effectiveness) have been adopted from the private sector by the present HE audit culture and new managerialism (audit itself being "a dividing practice" according to Shore and Wright, 2000, p. 61). Stephen Rowland remarks that the "separation between teachers and learners; the separation between academic staff and those who manage them; the split between teaching and research" is a deliberate action to threaten coherence (Rowland, 2002, p. 52). Fragmentation of knowledge workers from the rest of the HE professionals comes as a result of the ambivalent recognition of librarianship as a true profession by the rest of the academy.

Ethical professional responsibilities in a changing organisational world

In the public perception of what constitutes 'professionalism', ethical codes and trust are closely intertwined, as well as the assumption that the professional's self-interest comes second. Codes of ethics should be used to "create and sustain trust" and should serve to condemn "all those actions and circumstances in which the privileged position of practitioners is employed to generate profit beyond the value of the work that is performed" (Freidson, 2001, p. 215).

There is a range of factors that are seen as moral and ethical in each profession, which are consistent with philosophical definitions. CILIP has a responsibility to 'the public good' beyond the immediate limits of its membership. In the university, on the whole in the 'public sphere' (as conceptualised by Habermas, 1989), we find a commitment to rational debate and freedom of enquiry, an ethos which should also rule the professionalism of librarians. The CILIP code of professional practice lists the responsibilities that librarians have chiefly as employees[2], which are expected to be coterminous with those of the employers'. However, trust, rather than accountability, is argued to be the mainstay of professional practice (O'Neill, 2002), as current methods of performance accountability correlate with low trust (Avis, 2003), encouraging arbitrary and unprofessional choices.

The challenge to HE professionals (librarians included) hinges on not becoming accomplices in the administration of injustice, as professionals became in the Third Reich - where Berufsethos [professional ethos] was lost (Freidson, 2001, p. 131) and an unquestioning acceptance of the Nazi regime was shown by the universities (Barnett, 2000, p. 25). However, formal codes of ethics should not be mere manuals of administrative procedure, but should provide guidance for professional ethical issues - their disregard being, according to Freidson, another manifestation of an assault to professionalism (Freidson, 2001, p.216/17).

The librarian in a changing (he) world

A cursory glance at specialised HE library management literature shows that libraries are unwilling to innovate in any way that differs from the manner their academic peers are developing. "University A and University B [3] have taken different paths but arrived at the same destination" Moore states (Moore, 2003, p.151). Norry presents a case study, based on a University in the north of England, which identifies the following changes in staff roles, representative of the current trends affecting all universities - particularly those in the post-1992 sector:

- A tendency to conflate several roles in the role of 'librarian' (from different professions: ICT, information/library and technical media roles).
- Convergence of Library and IT departments.
- A desire for a strengthened partnership and liaison with academic staff - in terms of their pedagogic agenda, teaching delivery and learning models (Norry, 2003).

Interestingly, American HE libraries are preoccupied by the same set of issues that affect UK HE libraries: recruitment, education, and retention of librarians; the library as a place of intellectual stimulation and not a marginal service; impact of information technology on library services; preservation of digital resources; a change in the traditional library/publisher relationships; support of new users and reductions in HE funding, which impacts on library programmes, salaries, and resources, as well as the challenge of competition from other organisational units within the university (Hisle, 2002, p. 1-3). Some librarians, such as Brindley, Director of the British Library, while aware of these external and internal motives of change, are critical of the librarian's role as it has been constructed, and calls for diversity and dynamism in order to face these new challenges:

"Increasingly librarians will be judged as part of multi-skilled teams, as effective collaborators outside the comfortable box [...]. People, including librarians, are needed with higher level of skills and specialisms, and who are individuals willing to take risks, work in teams, be entrepreneurial, outgoing, manage projects, work quickly and approximately, and live in the messy world of the digital, where structured processes and procedures are often irrelevant. Not enough young people creating digital libraries see library school on their career path at present, or indeed libraries as somewhere exciting to work. Librarians have already missed many opportunities for engagement" (Brindley, 2001, p. 6).

In terms of defending our intellectual role, librarians as a group might not possess genuine monopoly over a body of specialized knowledge that could produce a 'knowledge gap', but individual librarians do, which they can use with considerable leverage over the user (Reeves, 1980, p. 137). If no professional librarians are used as information workers (in Freidson's terms, "if credentialism is not mandatory") library users must be trusted to make the right choices by themselves, a position that overestimates their discerning capabilities as consumers, especially in countries such as the US and the UK, where there is a high functional illiteracy rate (Freidson, 2001, p. 205).

The "assault" on professionalism

The professional issues addressed in this essay are not new. In the 3rd Annual Report of the CUKT dating from 31st January 1917 (alas, not a trendy French fashion label but the Carnegie United Kingdom Trust, which underwrote the activities of the Library Association) the following statement is found, and from which one can gauge how little the profession has changed its ways:

"The term 'librarian' is lightly used, and often applied to an official who is placed in charge of a collection of books, with very meagre knowledge of their contents and still less knowledge of the profession to which he purports to belong. This attitude on the part of some authorities results in a salary being paid which is totally inadequate for the responsibilities of the post. The inevitable result is that the profession as it stands today offers little or no attraction to a person with ambition, and who has a future to provide for." [quoted in Munford, 1976, p. 98].

In a new university in the north of England, most library staff roles were redesigned as part of the restructuring process in a review of academic services undertaken in 1996: "The most significant role change was that of Assistant Librarian to Information Adviser, demonstrating the deliberate move towards learner-centred support and hybridization" (Norry, 2003, p. 61). It is interesting how this author talks about the 'subject librarian' as an extinct species, and feels there is a need to explain what they were! (they are still 'alive and kicking' in Oxbridge, Russell Group universities, and in most so-called 'red brick' universities - as per D. Cook's classification of universities): "These were staff on academic grades who delivered teaching in information skills, were responsible for the acquisition and exploitation of resources in a subject area, and liaised with a specific school of study on information services and curriculum design and delivery" (Cook, 2005; Norry, 2003, p. 63). Retitling Librarians as Information Specialists, claims the author, sends a message of their information emphasis. In fact, what it sends a message of, is that their specialism is not respected anymore. The demotion of subject librarian to 'information specialist' is eclipsed by the importance given to the 'library managers', which are now in the limelight in HE in general: "Managers' roles were seen as vital to successful implementation of the change process and to making the new services work in practice" (Norry, 2003, p. 64).

Some regard specialisation in the library environment as counter-productive and undesirable. Smith refers to specialism[4] as a 'bad habit' picked up at library school and also as an internal unhealthy habit: "When professionals specialize to such a degree that they cannot understand each others roles, duties, or even terminologies, the effectiveness of the whole is diminished. [...] Academic librarians are typically more susceptible to the impairments of specialization than are public, school, or special librarians. [...]. Perhaps worse, library programs may recommend or even require that their students specialize in areas of librarianship. By actively limiting students in this way, programs inadvertently downgrade skills, undermining professional range and initial marketability. [...]A library staff composed of longtime, entrenched specialists also risks the development of territorialism [...]. Academic library patrons being at risk of poor service from a specialized environment." (Smith, 2003, p. 168-70).

Multi-tasking is favoured, as both the librarian and the manager of IT services are advised to form staff teams "to encourage multi-tasking and improve resilience to staff loss" (Buttle, 2004, p. 16). The argument is that qualified librarians can be 'saved' for more managerial positions (to the detriment of the reader's needs, as he/she no longer has access to qualified staff). The implication of this desirable 'multi-tasking' is that management do not want specialist librarians to mark their professional territory. Perhaps Freidson provides the ideology behind this anti-specialism position: "Since tasks and positions in firms are subject to change as productive ends and means change, commitment by workers to any particular job and body of knowledge and skill is obstructive and therefore undesirable" (Freidson, 2001, p. 109). If the professoriat is being proletarianized as a body by casualisation, part-time contracts, lack of tenure and other contemporary shifts (Readings, 1996, p. 1), this process is even more pronounced in the academic-related sector.

The user's centrality: fact or fiction?

The centrality of the library for students (as book prices soar) is rightly gauged by Barnett, despite his finding that the ritual [of visiting the library] is a strange affair, akin to a religious rite. "The significance that students attach to the library frequently emerges in student surveys: the opening hours and the availability of titles in both range and number (that is, duplicate copies) figure repeatedly as matters of concern to the students themselves. This pressure from the students is entirely proper." (Barnett, 1992, p. 39). In turn, students as customers of an institution should enjoy the same centrality. They have quite rightly become more demanding, seeking a return on their own financial investment in their education, with high expectations of the services and facilities provided. Reduced funding per student has put pressure on both schools of study and library services to reduce costs and increase efficiency. Against this background, some academic library managers question whether libraries are really user-directed. It is ironic that Sykes, as Head of Library Services at the London School of Economics, which has an outstanding Social Sciences library, should posit this tentative worry. "Where collection development and the custodial role of libraries still tended to hold sway right into the nineteen-eighties, now the users stand much more at centre stage" (Sykes, 2002, p. 2). The implication of the centrality of the user is a change in the role of librarians, who now identify themselves with a new species of 'hybrid learner-support professional', rather than a librarian or IT professional. Tellingly, this hybrid librarian "sees more relevance in achieving membership of the ILT than chartership of CILIP" (Moore, 2003, p. 132) which is a questionable result, as it erases professional identity. The desirable solution resides in the retention of our professional identity as librarians, without having to mutate into technicians or teachers. It should be remembered that an approach such as Total Quality Management (in which all employees at all levels concentrate on satisfying the customer) was tried and abandoned by corporations (Shore & Wright, 2000, p. 69), and rated in the USA as too expensive, bureaucratic and unreliable (Morley, 2004, p. 10). Hence the centrality of the user must be tempered by the demands of operational practice.

The librarian in an era of global challenge: is resistance futile?

How information and knowledge are accessed in a global era makes the idea of the library as solely serving a local academic community parochial, as Universities see themselves these days as 'world players'. For Deem the globalised world impacts on the local scene, as HE must search for new sources of finance (e.g. commercial research) and accept intervention in the form of audits, through government agencies such as the QAA or the Quality Assurance Agency (Deem, 2001).

If the university is in ruins[5], so is the library as a scholarly model. The seminar library idea (each collection mirroring the faculty structure, and serving as a teaching unit) is becoming obsolete, but it's preferable to a non-discrete sequence of books as it is mostly the case in the libraries of the new universities. The library is not immune to international influences: the Humboldtian ideal of the seminar library was imported from Germany in the early 1800s, when Prussia was a Mecca for students, as America is today. However, the university as a repository of national culture no longer exists, and has been replaced by the 'University of Excellence' - an adjective that has no intrinsic meaning. With globalisation (or Americanisation?) comes the decline of the nation-state, where the university becomes "a transnational bureaucratic corporation" (Readings, 1996, p. 3). Some call for the resistance of the "McDonaldization of the world" (Barnett, 2000, p. 103), and not just of the university or the library.

Historically, there has always been 'global challenges' that have affected the university as an institution and have had a knock-on effect on the library. Political changes in Germany in the early 1990's after the fall of the Berlin wall required a 'purge' of university staff who were seen as part of the regime in the former East Germany. The university personnel appointed to replace them (who might have been of equal competence) were "less easily identifiable as ideological agents of their state" (Readings, 1996, p. 41). The same 'unification' was carried out in 1992 between the libraries of the (East Berlin) State Library and its counterpart in West Berlin (built anew in the post-war period). A senior British Library Briton was appointed in 2002 (after 10 years of an uneasy partnership between the libraries) to introduce the brave new world of library reorganisation and 'new management'. He only lasted in his post one year, leaving "over difficulties of bringing together staff divided between two workplaces" and "over problems with organisational regulations that effectively prevented dismissals and reassignment of posts" (Burchard, 2003). It seems that in this case, the moves from an outsider (who, among other things, was thought not to understand the relationships between East and West, and wanted to turn the East German library into a museum) were resisted successfully, thus it may be a case that Resistance (in tandem with efficient labour law) is not futile.

Some authors glimpse other possibilities: Barnett sees the new post-modern university as an identity where multiple standards, multiple purposes, multiple knowledges and multiple consumers intersect: "a distributed university, taking up the agendas and discourses of its many clients, regionally and nationally and probably internationally" (Barnett, 2000, p. 21). This is a scenario, which can only but benefit the heterogeneous composition of the academic library, in terms of services offered and users' composition.

Potemkin villages: fabrication and performance in the academy

Higher Education is not immune to fabrications or a display of "performativity" which is in evidence particularly during the Research Assessment Exercise (RAE). Putting on an act when subjected to the gaze of authorities is not a new stratagem. Grigori Potemkin (1739-91) erected fake villages along Catherine of Russia's Crimean inspection tour, while local peasants were given new clothes and ordered to look cheerful. These "Potemkin tours" of factories were also popular during the Soviet era to impress visitors and government inspectors. Individuals are complicit in the construction and maintenance of fabrications, as they succumb to what Lyotard calls the "terrors of performativity" (Lyotard, 1984). One of Louise Morley's academic interviewees (a senior lecturer) declares (Morley, 2003, p. 165): "We've got better at playing games in the research assessment exercise [...]. And in the subject review, people have got much better at hiding things". Performativity seems therefore to produce opacity rather than transparency and accountability (Ball, 2003, p. 215). 'To keep systems volatile, slippery and opaque' is also singled out as an undeclared policy of UK Higher Education systems of accountability"(Shore & Wright, 1999, p. 569).

There is even a new language to describe roles and relationships. In Morley's study, one interviewee, a head of an education department who became an assessor for subject review, describes his experience (Morley, 2003, p. 115) as:

"Well, basically you had to learn to write what was called QUAAHILI - QAA prose ... it was basically inducting people into a set of behaviours which were acceptable within the sort of defined norms of QAA"

This linguistic behaviour is described as a form of ventriloquism, as the compliance culture of present HE produces "ventriloquizing apparatchiks" (Morley, 2003, p. 162).

Self-monitoring (via self appraisal systems, target-setting, output comparisons) is also set into motion, without a concomitant right to reply. This lack of accountability on the part of reviewers reminds some researchers of the one-way gaze of pornographic objectification (Morley, 2003, p. 157). Explicit coercion is not needed, as this "audit society" has internalised hopelessness, or a TINA effect - there is no alternative (Morley, 2003, p.165). Some authors claim that there is a psychic operation at work, which "requires the activation and exploitation of a range of feelings such as guilt, loyalty, desire, greed, shame, anxiety and responsibilisation, in the service of effectiveness and point-scoring" (Morley, 2004, p. 4). Foucault describes the process of normalization brought in by "disciplinary power" as the introduction of "the constraint of a conformity that must be achieved"(Foucault, 1979, p.182/3).

The implication for professional practice is that professionals are all players, to a great extent, in this panoptical game, concerned with watching and being watched, from the "innocent" open plan office layouts to sophisticated "dataveillance" of military systems (Levin et al, 2002). The library is not immune to being an auditable unit like the rest of the university. The language of "best practice" and of "forever improving" is not alien in this environment, however empty these concepts might be once analysed in our daily professional practice.

Feminisation of the he professions and patterns of employment

It has been accurately remarked that credentials and academic capital, once attained by women, mysteriously lose their value. Figures from the Higher Education Statistics Agency (HESA) show that the average female in academia will earn four to five years' less salary than an average male colleague for the same number of years worked. "Quite often my perception was that permanent jobs were going to men, and usually public school educated men who were similar to people making the decisions. There was no pressure to show criteria, or to explain the decision, or any of these things" (Female academic interviewed in Morley, 2003, p. 153).

Some view the RAE as beneficial for women, as the new mechanism offers more opportunities for advancement, away from the "old boy" network: "Because [...] the more prestigious institutions are the ones with the most hidden snobbery, but they need to achieve good results in terms of the RAE like everybody else. If you've got a lot of research these days they tend to pick you even if you're not their type." (Morley, 2003, p. 153).

The shift from masculinised management approaches to feminised ones is judged to be a favourable one (Deem, 2001, p. 11). However, we must consider the implications of a feminised workforce in libraries, which might have contributed to the negative image of the profession and its lack of influence in a male HE world. A graphic portrayal of women in HE describes them as "organizational Morlocks - the creatures who live underground and service the 'Upper World' people" (after H.G. Wells's subterranean creatures in The Time Machine, quoted in Morley, 2003, p. 157). Traditionally, most head librarians are women, but they do not have the institutional power that male heads of departments have (in 15 years of personal professional involvement, I have not witnessed any of the head librarians in HE rise to pro-provost or vice-chancellors).

Casualisation is a pattern of employment, which affects academic-related staff and supports workers the most. Johnson-Eilola (1997, p. 79) muses ironically over the "flexibility" of casual employment: "a growing number of these workers are temporary, to be given an even greater degree of individual freedom at the end of the project: freedom from employment, from heath insurance, from paid vacation work or sick leave, or from any of the other benefits normally accruing to workers" (Johnson-Eilola, 1997, p. 79). What Eilola omits to state, is that this casualisation practice is feminised: women are 33 % more likely than men to be employed on fixed-term contracts (The Guardian,1999, quoted in Morley 2003, p. 153). The implications of casualisation will be examined in the next section, as it has repercussions for our professional futures.

Conclusion: for librarians, what futures?

While attempting to frame alternative courses of action, we must bear in mind the conservative nature of librarianship and the refusal to change (after all, the internet was not invented by librarians) which ultimately might see their fragmentation or mutation if not their demise in the 21st century. In terms of staffing practices, it is very likely that we are already witnessing Freidson's description of "a two-tier professional system composed of permanent, relatively small elite corps of professionals who do research and set standards of performance [...] and a floating population of qualified practitioners who may be employed on a temporary and sometimes part-time basis" (Freidson, 2001, p. 209). The librarians will have to get used to a "horizontal" as opposed to a "vertical" career (mobility from one organisation to another in the labour market, rather up the hierarchy). Supercomplexity as presented by Barnett may also constitute a viable form of survival for the library: the incorporation of multiple frameworks of understanding and of action, in an organisational world with "no unifying notions which served another age"(Barnett, 2000, p. 21).

Although the present situation of jobs' reductions is lamentable, many subject specialists are not actively promoting the value of their work (e.g. their engagement with the curriculum via user services and maintenance of academic collections), which put their jobs and everyone else's in jeopardy. Librarians must acquire the ability to bring what Barnett calls a "reflexive biography" to life, in order to face to the present challenges addressed here, and others to come (Barnett, 2000, p. 159). If we are to survive as a profession, librarians should shift from a responsive model to a pro-active model, by being at the vanguard of innovation, and not be seen as mere implementers or book selectors, as there is no time or place for conformity.

REFERENCES

1. Avis, J.: "Rethinking trust in a performative culture: the case of education". *Journal of Education Policy*, 18(2003)3, 315-332.

2. Ball, S.: "The teacher's soul and the terrors of performativity". *Journal of Education Policy* 18(2003)2, 215-228.

3. Barnett, R.: *Improving Higher Education: Total Quality Care*. Buckingham : The Society for Research into Higher Education and Open University Press, 1992

4. Barnett, R.: *Higher Education: A Critical Business* Buckingham : Open University Press, 1997.

5. Barnett, R.: *Realizing the University in the age of supercomplexity*. Buckingham : The Society for Research into Higher Education and Open University Press, 2000.

6. Blake, N., R. Smith R. and P. Standish: *The Universities we need: Higher Education after Dearing*. London : Kogan Page, 1998.

7. Bourdieu, P.: *Homo Academicus*. Cambridge : Polity Press, 1984.

8. Brindley, L.: "What use are librarians (working in libraries)?". *Relay: the Journal of the University, College and Research Group,* 51(2001), 5-7.

9. Burchard, A.: "Graham Jefcoate leaves the Berlin State Library just one year after taking up office - in conflict over the future of the institution". *Der Tagesspiegel*, Berlin, 13.3.2003.

10. Buttle, A.: *Information Services Report*, Number 895/04. Southern Universities Management Services, 2004.

11. Chomsky, N.: *Chomsky on Miseducation*. Edited and introduced by Donaldo Macedo. Lanham, Maryland : Rowman & Littlefield Publishers, 2000.

12. CILIP. *When information goes wild... take cover or take control !*, 2004 (Leaflet detailing the routes and qualifications for an information professional).

13. Cook, D.: *Expansion and Innovation in British Higher Education, 1963-2003*. Innovation and Change in Higher and Professional Education Module, 18th of January 2005, School of Educational Foundation and Policy Studies, Institute of Education, University of London, 2005.

14. Deem, R.: "Globalisation, New Managerialism, Academic Capitalism and Entrepreneurialism in Universities: Is the local dimension still important?" *Comparative Education*, 37(2001)1, 7-20.

15. Foucault, M.: *Discipline and Punish*. New York : Vintage, 1979.

16. Freidson, E.: *Professional Powers: A study of the institutionalization of formal knowledge*. Chicago and London : University of Chicago Press, 1986.

17. Freidson, E.:*Professionalism: The Third Logic.* Cambridge : Polity Press, 2001.

18. Guile, D.: *We're all 'reflective practitioners now: a Utopian or utilitarian conception of HPE?* Innovation and Change in Higher and Professional Education Module, 8th of March 2005, School of Educational Foundation and Policy Studies, Institute of Education, University of London, 2005.

19. Habermas, J.:*The structural transformation of the public sphere: An inquiry into a category of bourgeois society*. Translated by Thomas Burger with the assistance of Frederick Lawrence. Cambridge : Polity Press, 1989.

20. Hisle, W. L.: "Top issues facing academic libraries: A report of the Focus on the Future Task Force". C&RL News, 63(2002)10. http://www.ala.org/ala/acrl/acrlpubs/crlnews/backissues2002/novmonth/topissuesfacing.htm

21. Information World Review. March 2005, issue 211. Bangor wake-up call, p. 9.

22. Johnson-Eilola, J.:*Nostalgic Angels: Rearticulating hypertext writing*. Norwood, NJ: Ablex Publishing, 1997.

23. Levin, T., U. Forhne and P. Weibel (eds): *CTRL [Space]: Rhetorics of Surveillance from Bentham to Big Brother*. Cambridge, Mass. : The MIT Press, 2002.

24. Lowe, R.: "Higher education". In: *A Century of Education*, ed. R. Aldrich. London : Routledge Falmer, 2002, pp. 75-92.

25. Lyotard, J.-F.: *The Postmodern Condition: a report on knowledge*. Manchester: Manchester University Press, 1984.

26. Moberly, W.: *Crisis in the University*. London: SCM Press, 1949.

27. Moore, K.: "Building new partnerships: changing institutional relations". In: *Centred on Learning: Academic Case Studies on Learning Centre Development*, ed. E. Oyston. Aldershot : Ashgate Press, 2003.

28. Morley, L.: "Quality and Equality: Intersections and Collisions". In: *Policies, politics and the future of lifelong learning*, ed. A. Hodgson. London : Kogan Page, 2000, pp. 85-96.

29. Morley, L.: Subjected to Review: Engendering Quality in Higher Education. *Journal of Education Policy*, 16(2001)5, 465-478.

30. Morley, L. *Quality and Power in Higher Education*. Buckingham : Society for Research into Higher Education and Open University Press, 2003.

31. Morley, L.: *Theorising Quality in Higher Education Bedford Way Paper 24.* London : Institute of Education, 2004.

32. Munford, W. A.: *A History of the Library Association: 1877-1977.* Old Woking, Surrey : Gresham Press, 1976.

33. Norry, J.: "The changing staff experience". In: *Centred on Learning: Academic Case Studies on Learning Centre Development,* ed. E. Oyston. Aldershot : Ashgate Press, 2003.

34. O'Neill, O. *Reith Lectures 2002. A Question of Trustt.* http://www.bbc.co.uk/radio4/reith2002/

35. O'Reilly, D., L. Cunningham and L. Lester (eds.): *Developing the Capable Practitioner.* London : Kogan Page, 1999..

36. Oyston, E. (ed.): *Centred on Learning: Academic Case Studies on Learning Centre Development.* Aldershot : Ashgate Press, 2003.

37. Raish, M. (ed.): *Musings, Meanderings and Monsters too: Essays on academic librarianship*. Lanham, Maryland : The Scarecrow Press Inc., 2003.

38. Readings, W.: *The University in Ruins.* Cambridge, MA : Harvard University Press, 1996.

39. Reeves, W. J.:*Librarians as professionals: The occupation's impact on library work arrangements.* Lexington, Mass : Lexington Books, 1980.

40. Rowland S.: "Overcoming Fragmentation in Professional Life: The Challenge for Academic Development". *Higher Education Quarterly*, 56(2002)1, 52-64.

41. Schon, D.: *Educating the Reflective Practitioner*. London : Josey-Bass, 1987.

42. Shore, C. and Wright, S.: "Audit culture and anthropology: Neo-liberalism in British Higher Education". *The Journal of the Royal Anthropological Institute*, 5(1999)4, 557-575.

43. Shore, C. and Wright, S.: "Coercive accountability: the rise of audit culture in Higher Education". In: *Audit Cultures: Anthropological Studies in Accountability, Ethics, and the Academy*, ed. M. Strathern. London : Routledge, 2000.

44. Smith, A. and Webster, F.: *The Postmodern University?* Buckingham : The Open University Press & Society for Research in Higher Education, 1997.

45. Smith, K.: "On Specialisation". In: *Musings, Meanderings and Monsters too: Essays on academic librarianship*, ed. Raish, M. Lanham, Maryland : The Scarecrow Press Inc., 2003.

46. Sykes, J.: "Are academic libraries really user facing?". *Relay: the Journal of the University, College and Research Group,* (2002)54, 1-2.

8
CHAPTER

Getting to Know Library Users' Needs — Experimental Ways to User-centred Library Innovation

Karen Harbo , Thomas Vibjerg Hansen

ASB Library, Business and Social Sciences, Aarhus University Library

ABSTRACT

'Meeting the User' is a programme committee under the Danish Electronic Research Library. As a development group at a national level we see our role as facilitating an innovative culture within academic libraries, focusing on users' needs and the way libraries meet them. In collaboration with a consultancy firm, the committee organized a travelling workshop in four cities in 2010. The workshop introduced practical ways for library staff to get to know their users' needs for services and was based on anthropological methods. The travelling workshop was part of a larger project called 'A Journey of Discovery in Danish Library User Land' (translated from Danish Brugerkaravanen), which also included a national thematic day, a blog and the publication of a method guide. There is an ongoing need for academic libraries to improve their services. One strategy is to become more aware of the users' needs. On the one hand we have libraries which give access to a lot of information, offer courses in information literacy and strive to be a part of the learning environment. On the other hand we are not always certain of the needs of our users. If libraries want to improve the way they serve their users' needs, they must innovate their services, facilities and courses by building upon what you could call 'user logic' and not upon classical 'library logic'. 'User logic' is that which is meaningful for the user instead of what is traditionally meaningful for a library. Finding out what the users' needs are, requires methods to study the users. In order to discover the shortest route from knowledge via idea to action, we looked for methods that can be employed by librarians or library information specialists. This article describes how 110 librarians and information specialists acquired such methods in four cities and in four days, by means of a workshop structured like a guided tour through the land of library users. The goal of the article is to give other libraries inspiration for ideas, concepts and concrete tools to study user behaviour and become more aware of the user's needs for service. The article contains a description of the above-mentioned methods, valuable experiences from the workshop, a presentation of the concept and concrete tools, discussion of the concept of user logic and library services, and the seven principles for human-centred innovation in relation to libraries, a short list of studies carried out by librarians and discussion of further perspectives.

Keywords: user-centered innovation, user needs, library innovation, anthropological methods, user logic, Brugerkaravanen

Introduction to 'Meeting the User'

The Danish Electronic Research Library (DEFF) is an organization which is co-funded by three Danish ministries: the Ministry of Science, Technology and Innovation; the Ministry of Education; and the Ministry of Culture. At the national level, DEFF collaborates with the institutions under the three ministries that co-fund DEFF, primarily libraries, academic institutions and secondary educational institutions. Internationally, DEFF collaborates with partners both in the Nordic region and the European region. The DEFF steering committee is responsible for DEFF's strategy and actual activities.

Four groups monitor DEFF's large domain of interest, which includes initiating projects (based on subsequent steering committee approval) and coordinating projects across the groups and with DEFF licenses. One of these groups, 'Meeting the User', worked as a development group at the national level from 2008 until 2011. The group's three main objectives were:

1. To facilitate an innovative culture within academic libraries focusing on users' needs and the way libraries meet them.

2. To initiate projects in the fields of information and knowledge creation, learning and communication at the junction between the physical and the virtual library environments.

3. To encourage sustainability of the project outcomes.

Innovation in Libraries

There is an ongoing need for academic libraries to improve their services, if only because of the considerable amounts of resources spent on library services and the need to ensure value for money. Libraries must take strategic action in order ensure that the services offered are the services needed. One strategy is to become more aware of the users' needs for services. On the one hand we have libraries which give access to a lot of information, offer courses in information literacy and strive to be a part of the learning environment. On the other hand we are not always certain of the needs of our students, teachers and researchers. What is traditionally meaningful for the library is not necessarily meaningful for the library user. There is always a risk that what libraries regard as the right answers to user needs, may be answers to questions that were never raised users in the first place.

In addition, users' needs may change imperceptibly over time and libraries must constantly take such changes into account. For example, today's students may still need guidance in using the enormous amount of e-based literarture, but at the same time they are, beyond dispute, advanced users of the latest information technology which library institutions, often conservative by nature, are only beginning to experiment with.

Offering services that are of no need to users is an expensive business. One ambition of the DEFF programme committee 'Meeting the user' was to challenge the state-of-the-art mindset of libraries and library staff by encouraging them to switch from what you could call classical 'library logic' to 'user logic'. Another ambition was that the new insights and knowledge gained by project participants could be transferred easily into their own library settings.

The Concept of User Logic and Library Services

Does a particular service appear to be logical and natural to a user, or does it appear to be logical to the library? This is the simple test question we ask ourselves when we innovate our library services. The next question is whether it should be either, both or neither.

Libraries support researchers and students in their research and learning and it is our job to allow them to benefit from the library's access to materials and services. In other words it is about the library's front end services. The question arises whether these should be based on the library's perspective of proper use of resources or on the user's perspective? In our terms we talk about library logic vs. user logic. The latter is our focus area!

User logic is about meaning. Are our services meaningful to our users from the users' point of view or from that of our libraries? Do our services appear to be logic to the users? Do they match the way users work?

The concept of 'user logic' is related to professor emeritus Carol C. Kulthau's description of the user's perspective on information retrieval and the paradigm shift from source or bibliographic oriented services to process-oriented library service (Kuhlthau, 2004). Briefly, Kulthau distinguishes between searching for sources (source oriented paradigm or library logic) and seeking meaning (process oriented paradigm or user logic). Sometimes library logic and user logic match, sometimes they do not. We have to remember this when we innovate our libraries and library services. We must also remind ourselves that the context is crucial: the kind of educational institution, the types of students and their learning background, the types of researchers and their topics, the educational culture and traditions, the social culture, etc.

A user logic point of departure means that we place the user in the centre, not the library. We do not focus on introducing the library's collections, but we focus on helping the user along in his or her research or study process from a holistic point of view. The library's collections are part of the services we offer, but they are by no means the only services (Figures 1 and 2).

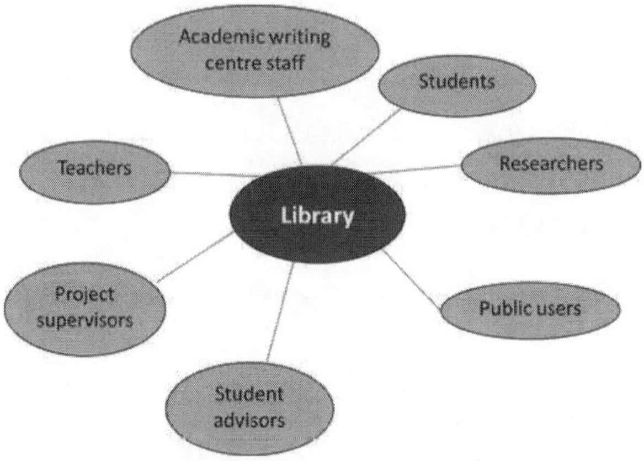

Figure 1. Library logic with the library in the centre.

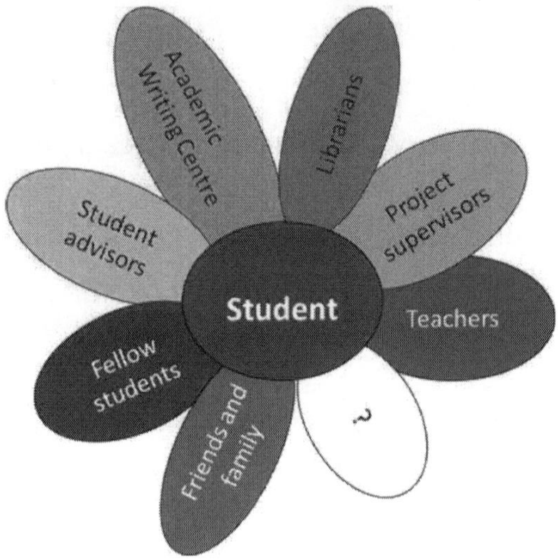

Figure 2. User logic with the user in the centre.

The classical library catalogue based on a classification system (e.g., UDK, Dewey) is a typical symbol of library logic. It forces the students and researchers to use the library in a particular way. Different libraries have different classification systems, even though they have the same kind of students!

The concept of user logic contrasts with library logic. It is not a well-defined concept, rather a concept we use for critical awareness when we want to innovate library services. We ask ourselves whether we want to innovate services, support, access, teaching, counseling etc. according to library logic or user logic. It is difficult to define a clear-cut rule about what is user logic and what is library logic. However, some symbols of library logic could be those shown in Figure 3:

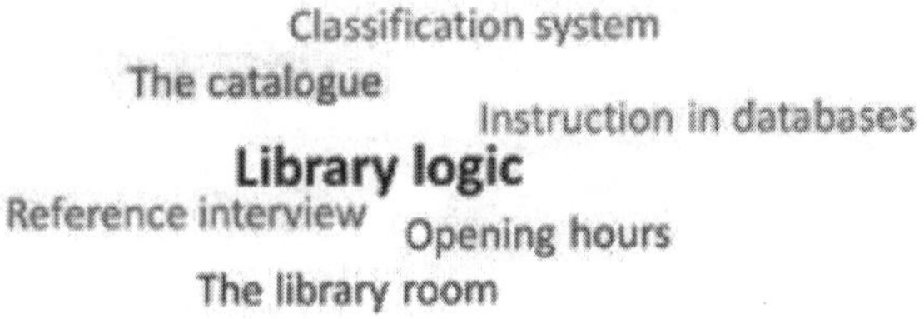

Figure 3.

Some examples of user logic test questions are:

1. Do you focus on the user's process or on the library's information resources in your library teaching, information counseling and other contacts with users?

2. Does your library reflect the users' needs, e.g., in terms of work spaces, facilities, access to materials, and resource or process-oriented help?

3. Do you understand how your users work or study?

4. Does the user require help to move along his studies or does the user need to know everything the library can offer?

5. Is the location of the service logical for the users? Should the services be available in the library where students and researchers work, or should they be available from the library web or another virtual place?

The methods, techniques and tools demonstrated in the project 'A Journey of Discovery in Danish Library User Land' help librarians to find answers to the above questions.

The Project 'A Journey of Discovery in Danish Library User Land'

In 2010 the programme committee 'Meeting the User', in close cooperation with an Aarhus-based consultancy firm named 'Changepilot' (John Lassen and Rebekka Høy Biegel), decided to launch a large and experimental development project on a national scale: A Journey of Discovery in Danish Library User Land. The expedition was termedBrugerkaravanen (which means: a caravan setting out for library users). TheBrugerkaravanen consisted of 110 library employees from academic libraries covering different levels of educational institutions. In sixteen mini-buses they visited 32 locations where users live their personal and professional lives in the cities of Copenhagen, Aarhus, Odense and Aalborg (Figure 4).

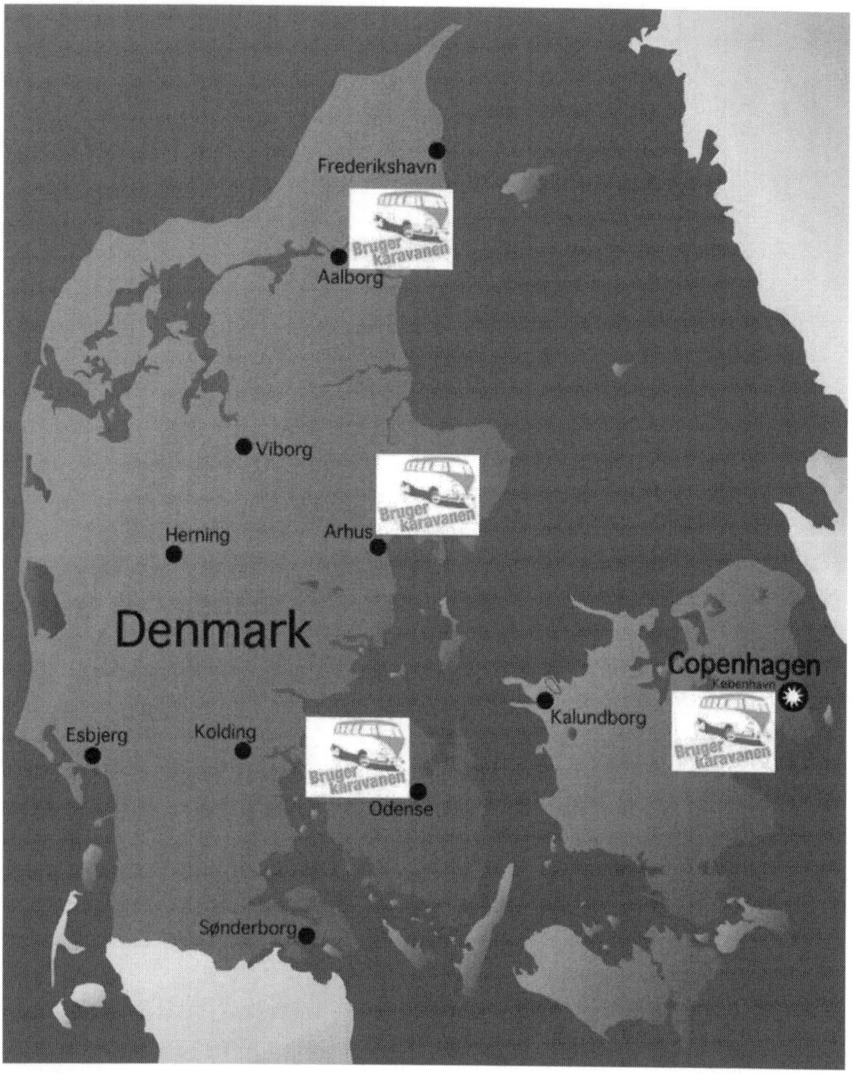

Figure 4.

In each of the cities meetings were arranged between the participants in the project and different categories of users. The project focused on four particular categories:

1. Master's students who already have completed higher education, are employed and have a demanding career.

2. Nursing students.

3. Exchange/international students.

4. Researchers in the field of arts and Humanities.

The consultancy firm 'Changepilot' that arranged the expedition to the four main cities of Denmark also arranged the meetings with these groups of users. In each city the participants were picked up in groups of eight by mini buses that took them to a 'base camp' for the tour of the day. By way of introduction the participants were shown a presentation about seven principles of human-centred innovation. After this they were introduced to qualitative methods and techniques inspired by the scientific fields of ethnography and anthropology: a set of user-centred tools for investigating users' needs. The presentations were intended as theoretical as well as practical background information to the programme of the day which included rides in minibuses to visit users that represented one of the above four categories.

The project had invited library staff to leave their (safe) libraries and go to where the users live, work, study, have coffee, listen to music, meet with friends and co-students. During these meetings with users in the field, the library staff were supported by expedition guides (employees from the consultancy firm). These helped the library staff apply the anthropological methods they had been introduced to, i.e., varied forms of interviews and observations. By interviewing and observing during the day, the library staff collected valuable data about the users. While collecting data they attempted to forget about libraries and classical 'library logic' and to focus on the people they met, not only as library users, but as ordinary people, living ordinary lives and doing ordinary work and or studies. By the end of the day, the data collected were analyzed, categorized and presented at the project base camps — still under the skillful guidance of the expedition guides.

After the four individual expeditions had taken place in the cities of Copenhagen, Aarhus, Odense and Aalborg, a final conference was held in Aarhus, to draw conclusions about the insights gained; these were presented as working questions by the participants by the end of each expedition. Many library staff participated in both events, but the final conference also attracted new participants. The conference took the work one step further and by the end of the day a publication was launched to document the methods and tools presented as well as a web site which also offered the participants a virtual space, a blog, to share knowledge on the subject in the future.

One participant of the Aarhus expedition shares her experiences

'Having signed up for the 2010 Journey of Discovery in Danish Library User Land in Aarhus, I had been told to meet in the hall of the central station at 9 o'clock. I had brought the little bright green suitcase that had been sent to me by the project, filled with items that had awoken my curiosity about the seminar on user-centred innovation which I was going to attend. While waiting I met other participants — easily recognized by the little bright green suitcases they all carried (Figure 5).

Figure 5.

'At 9 o'clock we were met by people of the project. A row of minibuses displaying bright signs with the name of the project waited for us in front of the station and we were taken, in groups of eight, to a place nearby that turned out to be the base-camp of this day of expedition.

'At the base camp we had coffee and were welcomed by the managers of the project and introduced to the programme of the day (Figure 6). First we were introduced to the principles of human centred innovation and to a set of qualitative methods and techniques — various kinds of interviews and observations — all of which were inspired by the scientific field of ethnography and anthropology. After being divided into groups of eight we were introduced to our team-leader — who would lead us through the practical exercises of the day, an expedition to library user land (Figure 6).

Figure 6.

'My group was to meet users from one of the four categories, i.e., master's students, and we were going to do research at their workplace and their home. Master's students are ambitious people who are studying to promote their careers, they often have a family and thus their days are full of duties. How do libraries create services that are meaningful to this category of students?

'The first stop was at the workplace of a financial advisor, who gave us a guided tour around his office. In this case we were told to use the method of observation in work context, to observe and take notes about the physical items and atmosphere of the office, and to take photos — if allowed —, and we were. After the guided tour we conducted abasic interview based on open questions and genuine curiosity and we got the opportunity to ask clarifying questions.

'The second stop was intended to be at the home of a media consultant, another master's student. But since he worked in Copenhagen most days of the week, it had not been possible to arrange such a meeting. Instead he had been asked to provide a photo diarywith pictures from his daily life and to write some text to each photo plus text that tied the photos together. After having examined the photo diary we did an interview with him in Copenhagen via mobile phone.

'Back at the base-camp we all had to categorize the data collected during observations and interviews (printed photos, video clips, observation notes, interview notes, and, in the case of taped interviews: transcripts, collected thoughts), by categorizing each segment under broader headlines. This phase was very crucial, we learned, as it is the careful preparation of data that lead to the final stage of the process: the analysis. Both phases should ideally be as broad and open as possible and conclusions not too easily and quickly drawn.

'When presenting our results, our group ended up by formulating three working questions, relevant for the creation of services for this category of users:

1. How do we relieve the master's students of their bad conscience?

2. How can libraries support talented employees to qualify for taking steps up the career ladder?

3. How can libraries work with information and services so that master's students' needs are met?

'These questions reflect the insights gained by actual meetings with the two master's students as ordinary people living ordinary lives — and not exclusively as library users.

'The Discovery Journey in Danish Library User Land thus gave me a range of concrete user-centred tools to choose from when I want to learn more about the needs of the users of my library. The methods and techniques are applicable in smaller as well as bigger library settings.' (Figure 7)

Figure 7.

Other User-centred Tools

The five other groups in the Aarhus expedition investigated users representing the three other categories and had the opportunity to try the other user-centred tools presented by the project.

Observations:

- Photo diary (described in text above)
- Observations in work context (described in text above)

- Think-aloud observations: library users are asked to solve certain problems while they think aloud during the execution of the assignment

Interviews:

- Basic interviews (described in text above)
- Value wheel: library users are interviewed in order to draw a picture of the extent to which their needs within a certain area are satisfied
- Interview at a location that means something to the user: library staff meet with library users on their home ground
- Small spontaneous interviews with users.
- Variations between interviews and observations:
- The meaning of physical surroundings: typical artifacts can be signs, boxes, furniture, doors, etc. — but may also include the way the light is set, the sound in a room and where it comes from, décor and much more
- Meaning of language — listening for metaphors: a useful tool to sharpen listening skills is to listen for the language. What are the terms and the metaphors used? Language tells a lot about the interviewed or observed person's approach to the world (Harbo and Biegel, 2010, p. 22 ff).

Seven Principles of Human-centred Innovation in Relation to Libraries

Many user investigations are carried out by using quantitative methods, collecting data through thoroughly designed questionnaires in order to get comparable data sets to analyze. However, quantitative investigations require big sets of data in order to be generally representative. Qualitative methods also require thorough preparation, but they can work with a smaller population of users, as qualitative methods do not claim generally representative results, but only results that are representative of the smaller categories of users actually investigated.

The methods used in this project belong to the latter category of user investigations, i.e., qualitative methods. Drawing on inspiration from the scientific field of ethnography and anthropology the project introduced the actual practicability of a set of abstract methods: a range of user- centred tools which could realistically be implemented even in small library environments — a criterion of success which had been defined by the project.

Teaming up with 'Changepilot' gave us a methodological basis for an open-minded, anthropological observing and listening approach. To make the approach practical in our everyday library work, 'Changepilot' formulated seven qualitative methodological principles about human-centred innovation which enable us to focus on user logic (Harbo and Biegel, 2010, p. 8 ff).

A short description:

1. Users run nothing

We prefer to use the term 'human-centred innovation' instead of 'user-driven innovation'. For most people it is difficult to put new words to what they need. Instead we try to uncover what drives people, what they dream of and how they live their lives.

2. Remove the blinkers

In areas where we basically are the experts (e.g., the library) our own knowledge may be the biggest obstacle that keeps us from seeing new angles. We must mentally remove our blinkers and put on new glasses to see things with new eyes.

3. Focus on the question

When working on human development one of the pitfalls is that we begin to search for solutions at the same moment we see a challenge. It is important to focus on the questions and to refine the question or to find more questions in order to be sure that we have formulated a question which is important for the users.

4. Meet ordinary people

We must meet and know ordinary people to find out what really matters to them to come up with exceptional solutions.

5. Focus on context

The decision to go to the library is taken before the visit takes place and the value of the visit is realized afterwards in the user's everyday life. Therefore we need to focus on the whole context surrounding the visit, including what happens before and after the visit.

6. Improve people's lives

We develop and innovate to improve people's lives. Nothing less!

7. Don't go to the zoo, go to the jungle

Human-centred innovation is forcing us out there where we meet the user in his or her own element, not in ours!

What does that mean for the librarian at work? Librarians who like to be in control of a situation and find proper answers are challenged by the human-centred approach to be open-minded and truly observe the user in his or her needs. The seven principles give answers to how and why to use this approach. How: principle 2, 3, 4, 5 and 7. Why: principle 1 and 6.

Perhaps the librarian experiences methodological pressure when using this approach. In that case the librarian must ask herself what is practically possible within her working context. Every new insight is valuable and is good training for the librarian using anthropological methods. So the use of the seven principles is scalable in relation to the librarian's working conditions. It should be possible to accomplish human-centred innovation if you are the only librarian in a small library or if you are a member of a team of librarians at a large university library.

At the same time living up to the seven principles keeps the 'anthropological librarian' on the right human-centred track.

'Users run nothing' means that the librarian should not only ask the user 'what do you need?', but observe and perhaps instead ask 'what is important for you?' in relation to the academic task, and in relation to the user's study life in general.

'Remove the blinkers' is very difficult for a librarian or any professional exploring his own area, but patience and not coming up with quick answers are ways to start seeing new things. A lot of our professional work is based on routine and common library knowledge, but trying to be a student or a researcher could provide us with new glasses.

In practice the librarian should dare to be an 'anthropologist' and go outside the library to follow a researcher or a group of students, e.g., by being introduced to the researcher's work process and teaching preparation, or by following study group meetings and courses, or lunch breaks by students.

'Focus on the question' is again a challenge to a librarian's need for control and the wish to give proper answers quickly. Normally a user will ask and the librarian will come up with an answer. Changing the roles will be difficult.

'Meet ordinary people' should not be a challenge to a librarian working with front end services. However, it is important to understand users not only as library users.

'Improve people's lives' could be specified as a university library's objective to improve not only education and research but also the lives of students and researchers.

'Don't go to the zoo, go to the jungle' means meeting the users on their home ground, where they work, study, have lunch and coffee breaks, live.

All in all human-centred innovation means a more proactive, process-oriented, observing and anthropological type of librarian who is ready to seek meaning and whose objective is to improve users' lives. The whole 'field trip to user land'-concept is a step towards giving librarians the competences to do that.

However, keep in mind that the concept can be implemented on a small scale as well as on a large scale, and that a pragmatic attitude is necessary.

Conclusions, Sustainability and Further Perspectives

The DEFF project A Journey of Discovery in Danish Library User Land demonstrated concrete approaches and methods which in the future will make Danish librarians capable of perceiving library users in a new light. To perceive the library user as a person with a complex set of needs is very different from the way in which libraries have traditionally perceived their users. Complexity is here an important and meaningful element and insights in users' own understanding of their lives can contribute to radical changes in the way in which library services are presented.

However, the acknowledgement of the fact that the interpretation of users' behaviour is not necessarily the same as users' own interpretations is just as important. In this context the user-driven concept is substituted by the user-centred concept, which allows for inclusion of libraries' analysis of the field work. By implementing the range of user-centred tools presented by the project, libraries have a real possibility to start transforming the mindsets of librarians with regard to library users.

The project revealed completely new ways for us, library staff, to meet library users. We realized that in their lives (representing the user logic), libraries did not mean as much as we, library staff, think. Thus, the library staff participating in the expedition obtained new and valuable insights into the land of library users. The qualitative methods and techniques that they have been introduced to by the guides of the expedition have proven to be very useful and easily transferable when the staff returned to their own libraries.

Examples of New Ways of Librarians Meeting Library Users — Inspired by the Project

- Observation of behaviour of patrons using a web application which tells them where a book should be placed in order to test if the application worked well in a certain context. The point was that saying is one thing, doing is another! This is why observation was preferred to interviewing.
- A small exploration of the understanding of the words 'community' and 'motivation' in relation to the question 'what does a good day of study look like'? University students at the Department of Social Sciences were asked this question by librarians using tools from 'A Field Trip to Library User Land'; they focused especially on the 'value wheel'.
- Another small investigation of the reason why some library users hesitate to ask the librarian/information counselor for help or, alternatively, started the conversation with 'I am sorry to disturb you'. Library staff were worried that they looked too busy, unfriendly or unapproachable. They observed library users and during one day they interviewed library users who started their approach with 'I am sorry to disturb'. Seven small interviews were done.

- A library team wanted to explore the concept of service by observing staff and customer interactions and by interviewing people using vox pop at 1) a combined swimming pool and library, 2) a shopping mall, 3) outside a tea and coffee shop, 4) a department store, 5) a MacDonald's, and 6) Ikea. What should their library service look like?
- Observations of library users combined with questionnaires sent to 180 students asking them whether the library should keep the collection of reference books or replace it with magazines about art, literature and music.

FURTHER PERSPECTIVES

Do we always want a library to be based on user logic (process oriented) innovation? We think it will make a lot of sense considering the fact that library services are targeted at users. Examples are the library catalogue, the classification system, the design of the library, the position of meetings point, information counseling services, etc.

We should also ask ourselves in which situations we prefer library logic (product oriented) innovation. In some cases, it is necessary to get a specific message through, e.g., how to navigate in a specific database.

According to the situation, either orientation can attribute to improve library users' lives.

Every librarian should be able to implement human-centred innovation in her library. The challenge for us is to further a human-centred innovation culture among librarians and to spread the word that it can be done in many different shapes and sizes. In our pragmatic understanding it is better to focus on the outcome than necessarily having one hundred per cent scientifically based evidence. If we succeed, we believe libraries and librarians will become flexible in adjusting to their users' needs.

With the easy access to information through, e.g., Google, Wikipedia and other websites the users' need for libraries providing access to information is perhaps decreasing. On the other hand the need of information counseling is increasing as access to enormous quantities of information resources requires information-literate users. Information counseling of users requires understanding of the context in which the users find themselves. The counseling librarian needs to know as much about the users' way of working, studying and living, as he or she knows about running a library in an efficient way.

One way of getting to know the users' needs is by applying the methods and techniques demonstrated in the project 'A Journey of Discovery in Danish Library User Land'. We recommend that a shift of focus from library logic to user logic be discussed in library environments as well as at academies of library and information science.

The project 'A Journey of Discovery in Danish Library User Land', its workshop concept, its how-to-do-manual containing concrete tools, and its website to share experiences and discussions, give us a practical fundament. Now it is up to us to do it!

REFERENCE

1. Kuhlthau, Carol C. (2004): Seeking meaning — a process approach to library and information services, Libraries Unlimited.

2. Harbo, Karen and Biegel, Rebekka (2010): A Field Trip to Library User Land. Deff — Denmark's Electronic Research Library and Changepilot.

3. Deff — Denmark's Electronic Research Library: www.deff.dk.

4. Changepilot: www.changepilot.dk.

9

CHAPTER

Co-Teaching in Information Literacy During Work Placements: The Librarian's Role

Anett Kristin Kolstad

Oslo and Akershus University College of Applied Sciences, NO

ABSTRACT

Introduction: This article presents findings from a development project designed to improve Information Literacy (IL) education by linking it to nursing students' application of evidence-based practice during work placements. The Learning Centre at Oslo University College in Norway aimed to find out, what interactions and conditions influenced the librarian's role in the context of interdisciplinary supervision of nursing students' assignments and the development of a common supervision culture.

Research Design: The librarian involved in the initiative had a dual role as both researcher and practitioner. Besides direct experiences from being a participant in the project, the empirical material consists of observation notes, meeting notes, presentations, guideline in IL, e-mails, an interview, 285 comments from students, nurse educators, nurse supervisors and librarians, through blog and the Wiki Learning Management System (LMS). The evidence on which this article is based involved analysis of both oral and written communication, framed in practice-oriented approaches.

Findings: The conditions influencing the changing position of the librarian from an 'outsider' to becoming a member of the community of practice with common structures for the collaboration and supervision included meetings, a blog and a Wiki. Multidisciplinary, complementary skills in the supervision teams provided improved quality of supervision and led to a new joint supervision community.

Conclusion: Together with the nurse educators and the nurse supervisors, the librarian was able to be a partner and a catalyst of IL workplace learning. The LMS was a decisive tool in the process. However, to be successful, this changed role for the librarian may require further pedagogical training and expertise in the development of interdisciplinary educational supervision.

Keywords: academic library, university, workplace information literacy, co-teaching

1. INTRODUCTION

Nurse educators and librarians at Oslo University College,1 in Norway, were keen to ensure that students learn information literacy (IL) to be able to use evidence-based practices (EBPs) during their training. This is seen as important for students to stay abreast of current developments and apply them in the workplace. The experiences of the nurse educators and librarians showed that it was a challenge to prepare students for the transition from education to the workplace. In 2009, the Learning Centre2 was invited to participate in supervising 50 nursing students, in one of the largest work placement initiatives in Oslo. The guidance was to be within the framework of the*Langerud model: a learning model for value and evidence-based nursing in nursing homes providing elderly care* (2009–2012). The Learning Centre established a sub-project of the Langerud model, called *Leeway for information seeking* to try out new forms of guidance in IL. This article focuses on the interactions and conditions that shaped the librarian's role in interdisciplinary supervision of student's workplace assignments and the development of a common supervision culture.

1.1. Some Key Issues Highlighted by the Literature Review

Evidence based practice (EBP) requires that decisions about health care are based on the best available, current, valid and relevant evidence (**Dawes et al., 2005; Sackett, Rosenberg, Gray, Haynes, & Richardson, 1996**). These decisions should be made by those receiving care, informed by the tacit and explicit knowledge of those providing care, within the context of available resources. The implementation of EBP is a complex process that requires knowledge and skills to ensure practice change is successful and sustainable. IL competencies lay the foundation for EBP (**Cheeseman, 2013**). IL is defined as a set of abilities requiring individuals to recognise when information is needed and to have the ability to locate, evaluate and use effectively the required information (**Association of College and Research Libraries (ACRL), 2000**). More recently the ACRL put together IL competency standards specifically designed for nursing (**ACRL, 2013**). IL competencies are considered as important in the contemporary environment of rapid technological change. In the context of nursing, it is essential for all graduating and practising nurses to base their practices on current, high-quality research evidence (**Cheeseman, 2013; Forster, 2013**).

In this article, learning IL is considered from a socio-cultural theoretical perspective; where knowledge is seen to develop through situated interaction in authentic contexts. Learning is seen as taking place through communicative interaction between people and tools in the discursive practice of the work placement. In this study, the understanding of IL is tied to the work placement. A socio-cultural perspective has been applied by**Limberg and Sundin (2006)** to explore IL in educational settings. The sociocultural approach considers IL as a set of abilities to seek and use information in purposeful ways related to the task, situation and context where it happens, in this case, the work placement. It is worth underlining that this view, implying IL as shaped through social interaction and varying between situations and contexts, differs fundamentally from the ACRL standards focussing on general individual competences. **Limberg and Sundin (2006)** particularly point out the importance of teaching content and context related to user perspectives. The socio-technical practice approach sees the IL movement as being primarily about enabling groups and communities to cultivate existing information strategies and supporting them in their interactions with information technologies (**Tuominen, Savolainen, & Talja, 2005**). In the article, the interplay between the use of the LMS as a collaborative tool and IL teaching at the work placement is highlighted. Several researchers have applied socio-cultural perspectives in exploring how information seeking is negotiated as part of the workplace learning process (**Bonner & Lloyd, 2011**; **Lloyd, 2013**; **Sundin, 2003**). **Lloyd (2013)** has investigated nurses' information seeking in their first year in the workplace and presents a 'people-in-practice' approach, shifting away from skills towards social practices. Lloyd identified how information emerges. Ways of knowing are seen as the collaborative enabling of connection with a range of knowledge and ways of accessing it (pp. 222–223).

Lloyd also highlights the role of co-participation in the community in shaping the production, reproduction and circulation of knowledge, including knowledge about the appropriateness of information skills in relation to the context. She emphasises that: "...without the catalyst of IL, workplace learning cannot occur. The two are entwined in joint enterprise". She raises important issues about the role librarians should play in education in workplace IL and examines how librarians can build information resilient workers, who can deal with rapidly changing information environments. She argues for the need to develop IL pedagogy that takes into account the transition from learning to the realities of the working environment. Sundin's research focuses on how nurses at a hospital shaped their identities through learning and applying information seeking. His findings indicate that nurses want to learn information seeking in their hospital departments, rather than via courses in the library, even if there is a library at the hospital. In the library, information seeking seemed too abstract. The nurses felt that the library and librarians could have a mediating role in supporting them (pp. 196–197).

There are several researchers, who have investigated the role of librarians in different IL training collaboration models with faculties of higher education; examining how IL research in workplaces affects the librarian's role (**Brasley, 2008; Mounce, 2010;Vassilakaki & Moniarou-Papaconstantinou, 2015**). Literature reviews show that there are numerous studies about how librarians and faculty work in the planning phase to integrate IL instruction into the faculties' courses and to prepare librarians to teach students on campus. It is an important part of the process to be integrated into the curriculum, but as highlighted by **Shank, Bell and Zabel (2011)**: "...librarians can and should be integral educational partners as well as a catalyst for students' knowledge, enrichment and inquiry". This is consistent with Lloyd's thinking and also presupposes collaborative teaching partnerships between librarians and academics. Four key points are identified for establishing effective collaboration within a faculty and library; 1) a shared and understood goal, 2) mutual respect, tolerance and trust, 3) competence for the task at hand from each of the partners and 4) ongoing communication (**Ivey, 2003**). Brasley outlines some interesting examples of collaborative models where the librarians are integral educational partners from the planning stage of the teaching sessions to the completion. One of the studies mentioned, is UCLA's Freshman Clusters, which uses a learning community approach to introduce students to a specific topic (**Lindholm, 2007**). Librarians at

UCLA's Library were involved as liasons to the clusters in the interdisciplinary courses from the nascent stages of the programme that was initiated already in 1997–1998. Faculty members, teaching fellows and librarians collaborated on the development of content in Web pages, design of IL courses, critical thinking activities, learning outcomes for the students and participated in electronic discussion board conversations between students and faculty. Another relevant model is Sonoma State University, Freshman Year Experience (FYE) IL programme. Discipline and library faculty worked cohesively to integrate information competencies into their English 101 courses. The successful partnership operated under the premise that teams should be flexible and emphasise each person's area of competence. One FYE faculty member stated: "I describe the process of teaching alongside an IL specialist as synergetic—producing end findings that are more than the sum of the teaching parts" (**Brodsky & Bower, 2005**). Another example mentioned is the Ball State Nursing Programme that had already started to create Web-based modules in a course management system in 2001 (**Florea, 2008**). It is very similar to the way in which the Learning Management Systems (LMSs) were used in the project *Leeway for information seeking* (**Kolstad, 2009**). In Florea's study, complementary expertise and sharing knowledge was emphasised as important for both digital and face to face tuition. The common aim was: "to ensure that all nursing students graduated with extensive IL expertise".

Tilley (2011) also emphasises that the librarian should be an integrated partner throughout the programme in a manner of 'co-agency' and 'co-learning'. It is necessary to pool expertise from all the participants to teach information skills for successful engagement. Tilley states that: "Co-agency is based on making connections, about sharing responsibility in the learning environment. The principle of co-learning is based on pedagogically appropriate learning where learners are actively involved in the learning process" (**Tilley, 2011**, p. 135).

Most of the IL research studies take place on campus or as workplace studies, as mentioned above. There is currently little research about the potential role of the librarian in the transition between the campus and the workplace; for example, participating in joint supervision teams during student's work placements. There is also little research on how a LMS may function as a supportive tool in joint supervision.

1.2. Aim

The aim of the research described in this article was to find out what interactions and conditions shaped the librarian's role in interdisciplinary supervision of student's workplace assignments and the development of a common supervision culture.

The key questions addressed by the project were:

- How did the nurse educators, nurse supervisors and librarians communicate when they supervised the students?
- What were key elements, such as tools and topics, in their communication?
- In what ways did the librarian develop a new pedagogical role by co-teaching in work placements with regard to her position related to nurse educators and nurse supervisors?

The article begins by introducing the study of joint supervision as part of *Leeway for information seeking* and describes the methods used to analyse the findings based on Wenger's understanding of learning in communities of practice (**Wenger, 1998**). The findings are analysed and presented according to the principles of a shared repertoire, joint enterprise and mutual engagement (**Wenger, 1998**, p. 95). Finally, the findings are discussed in the context of recent research relating to how the librarian can become a 'partner' during educational work placements.

2. THE STUDY

Leeway for information seeking became a part of the *Langerud model* in 2009, because the Learning Centre was invited by the nursing faculty to participate with a librarian in the project. The project was designed to ensure that students learn IL in order to be able to use EBPs during their training. Joint supervision from nurse educators, nurse supervisors and librarians was set up as a means to reach the goal. In the Langerud model the hypothesis was that interdisciplinary guidance in work placements, together with a stronger focus on the evidence-based framework would provide improved learning outcomes for the students. This was seen as an improved alternative to learning IL inside a classroom on campus. This was a new approach to teaching for the Learning Centre.

2.1. The IL Teaching on Campus

Prior to this project, IL training was performed on campus and consisted of a two hour training session in a classroom in the first year of the nursing Bachelor´s degree. The content was planned in collaboration with the nurse educators and the focus was on the students' tasks. The students were required to use authentic problems in their own tasks and find one academic or research article in the Nordic medical database Svemed+ and the international nursing database CINAHL. The session gave an introduction to information seeking using the PICOT format (**Melnyk et al., 2010**). Inquiries in this format take into account patient population (P), intervention (I), comparison (C), outcome (O) and time (T). The format provides an efficient framework for information seeking in medical science databases and is a part of the seven steps of evidence-based practice. This is described fully in section 3, because it became a requirement in the*Langerud model*.

2.2. The Project Leeway for Information Seeking

The project manager of *Leeway for information seeking* was a librarian as is the author of this article. She was also one of the supervisors. The whole project was managed through a number of groups with various responsibilities and interests. The *steering* group included the head of the workplace and the head of nursing education. The *reference*group consisted of the eight heads of different nursing departments. The *programme*group consisted of the leader of the Langerud model programme, who was also one of the nurse educators on the supervision team. In addition, it consisted of two or three nurse educators, the project manager of Leeway for information seeking, who was also one of the librarians in the supervision team and one professional development nurse. The development nurse had a coordinating role for nursing supervisors and was responsible for promoting their contribution in the collaborative supervision of the LMS.

The composition of the supervision team varied between one or two librarians, two or four nurse educators and about 50 nurse supervisors. This was slightly different in each practice period, depending on available human resources. The team supervised about 50 nursing undergraduates in each work practice period. There were a total of eight practice periods, which involved about 400 students during the project period. The students were divided into eight learning groups, one learning group per department. Over a period of eight weeks, tuition took place through the LMS, responding to the students needs and at weekly one hour meetings with the learning group. The librarian particularly participated in the joint supervision sessions with one of the nurse educators and the nurse supervisors in two departments during two practice periods. However, she also supported the other six learning groups when the nurse educators requested her inputs. At the end of their practice period the students were required to give an assignment presentation to the department that required information seeking in scientific databases to find research articles. The teaching team supervised students while they worked on their assignments. Students were asked to write a paper about an issue that was relevant for the department of the work placement and their own studies. They could also choose to write a paper based on a specific group of patients. The joint supervision of the student assignments is illustrated as one of several communities of practice in Figure 1.

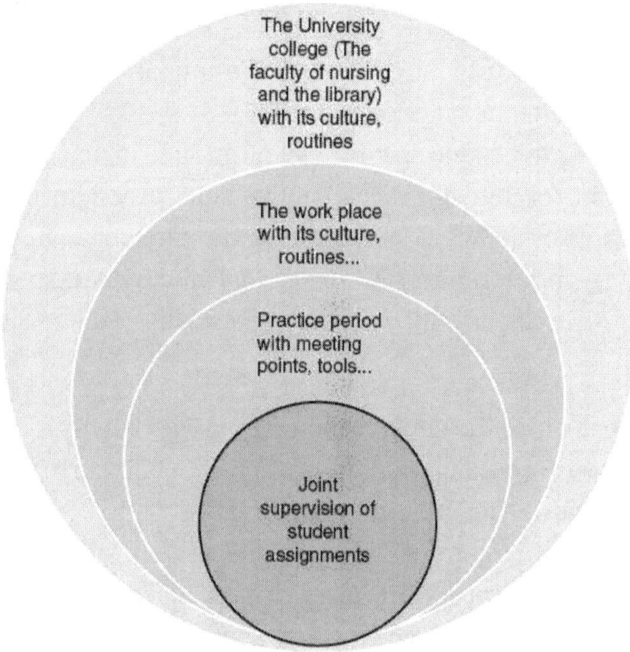

Figure 1. Community of practice: Joint supervision of student assignments.

The students were given guidance according to the seven steps of evidence-based practice (**Melnyk et al., 2010**):

- 0. Cultivate a spirit of inquiry

The students were encouraged to be inquisitive about how practices work and to ask critical questions in the following manner:

- 1. Asking clinical questions in the PICOT format

  Inquiries in this format take into account patient population (P), intervention (I), comparison (C), outcome (O) and time (T). The format provides an efficient framework for information seeking in medical science databases.

- 2. Searching for the best evidence

  The search is streamlined when questions are asked in the PICOT format. This helps to identify key words or phrases that, when entered successively and then combined, expedite the location of relevant articles in databases such as MEDLINE or CINAHL.

- 3. Critically appraising the evidence

  The students were asked to consider whether the findings were valid and if they were relevant for the patient.

- 4. Integrating the evidence with clinical expertise and patient preferences and values

  At this stage, the students were required to present their assignment to their related department and the nurse educator, nurse supervisor and the librarian. They had to discuss how they searched for the best evidence, their findings and considerations. The last two steps were not considered, because they were not naturally tasks for the students:

- 5. Evaluating the outcomes of the practice decisions or changes, based on evidence

- 6. Disseminating EBP findings.

Sundin distinguishes between clinically oriented information seeking and academic oriented information seeking (**Sundin, 2003**, pp. 173–175). A cross-over between these two is the idea behind EBP in the assignment that the students were required to carry out. Clinically oriented, in the sense that it should be relevant to the workplace and the students should be able to justify it in front of the department in the workplace. It is also an academic exercise where all the procedures should be in accordance with the seven steps, using different kinds of literature. They also had to use up-to-date research and the American Psychological Association style for references (**American Psychological Association (APA), 2010**).

3. RESEARCH APPROACH AND DESIGN OF STUDY

The librarian, who was the project manager of the Leeway for information seeking, had a dual role being both researcher and practitioner simultaneously. The influence was of course subjective, a fact that may have affected the results. By participating in different supervision teams in the workplace, the librarian as a researcher gained insights into the needs of the students and acquired useful experiences and understanding through communication, negotiation and co-teaching with the nurse educators and nurse supervisors. However, she may have also become so involved that it could be difficult to be critical.

3.1. Research Approach

The focus in this article is on how the supervisors used language to support each other with their own competence to give the students the best possible guidance. The sociocultural perspective underlines the importance of interactions and communication with other people in the community for what is being learned and how. Language is at the same time a collective, interactive and individual sociocultural tool (**Säljö,**

2005, p. 87). By combining a variety of insights and experiences in the teams, supervision aimed at better quality. The comments from supervisors were saved in the LMS and were accessible for all the participants. In this way, knowledge was distributed.

In combination with textual analysis, the empirical data was analysed by using Wenger's three conditions for a community practice: joint enterprise, mutual engagement and a shared repertoire. The analysis aimed to establish whether a new joint community of practice emerged over time via the collaborative student supervision. Wenger defines a community of practice as a group of people, who share a common commitment to something that they are engaged in and so learn from each other how they could do better through regular interaction. According to Wenger, this process of participating in various communities of practice is linked to the shaping of identity. This involves the process of negotiation in practice and engagement with it. Group identity develops through inclusion, exclusion, commitment, unity, trust, recognition and inspiration.

A shared repertoire gives participants common reference points, recognition and opportunity to reuse the repertoire in new situations (**Wenger, 1998**, pp. 101–102). This relates to what the community has produced in terms of routines, words, concepts, tools and ways of doing things. It encompasses the discourse that occurs, the manner in which they express forms of membership and identity as members. In the study it was important to find out if a common repertoire was developed, despite the members having different vantage points.

According to Wenger, negotiation of joint enterprise implies ways of working with and solving tasks by participants in the community of practice. He states that multiple memberships and multiple commitments are useful to achieve synergies. Everyone has something to contribute to determine the conditions, resources and requirements that will apply in a community of practice. Participants must find a way to do this together. Objects and focus points are shaped as meaningful negotiations to organise around (**Wenger, 1998**, p. 84) and common tasks that all are mutually responsible for (**Wenger, 1998**, pp. 95–96), for example, the programme for the practical period, agreement in how to supervise in information seeking, to use tools such as blog and wiki.

Mutual commitment defines a community (**Wenger, 1998**). This assumes that the community can communicate and collaborate. In the workplace, one can facilitate the engagement of more people with further or different expertise if required. In this study, engagement of different expertise was facilitated by the structure of various meetings and the use of the LMS. The value of different factors facilitating mutual engagement comes out in the presentation of findings.

3.2. Design of Study

As already mentioned, the author of this article was involved as an active participant in the project studied. This means that the empirical material was gradually collected both formally and informally through participation during the whole project period 2009–2012. The material thus consists of the author's experiences as a participant, combined with observation notes, meeting notes, course presentations, e-mail communications, written guideline in IL instruction in for the supervisors, statements about the use of the LMS, which were made during IL training, an interview with the project manager of the Langerud model and a total of 285 comments from students, nurses, nurse educators and librarians (see section 2.2).

Source quotations are coded as PM for project manager, NE1-4/5 for nurse educators, NS1-15/50 for nurse supervisors and L1-3/3 for librarians and the practical periods are coded as S2010(2), S2011(1), S2011(2), S2012(1), A2010, A2011for practical period in spring(S)/autumn(A) and year.

This material was analysed to investigate and identify communicative patterns to find main tendencies in how the supervisors "spoke" when they answered the students and how they engaged with each others' supervision. A particular focus was directed at the position of the librarian in the communicative patterns. The material on which this article is based combines qualitative text analysis of oral communication in the learning groups and written communications from the LMS. Elements of conversation analysis seemed useful for both purposes and ideas to do linguistic text analysis were picked from**Rapley (2007)**.

The expressions were analysed to find categorised in the following way:

- Personal and uninhibited communication
- A culture of sharing
- Towards a common vocabulary
- Acknowledgement of each others competences
- Support to the students

Owing to insufficient documentation, observation notes of communications in meetings are not as complete and thorough as the analysis of the comments on the LMS. There is therefore a strong emphasis on the findings of the analysis of the text on the LMS. Notes, e-mails, presentations and observations are nevertheless analysed, in order to get a richer picture of the context surrounding the discussions.

4. RESULTS

The findings are presented in such a way as to give an overview of the interactions and conditions that appear to have had implications for the librarian's role as a partner in joint collaborative supervision. The findings are analysed and presented according to the principles of a shared repertoire, joint enterprise and mutual engagement (**Wenger, 1998**, p. 95).

4.1. A Shared Repertoire

This section highlights examples from conversation analysis of sequences in the LMS and the student departmental presentations, to illustrate how the supervisors used their expertise. This will be discussed in terms of the seven steps of evidence-based practice (**Melnyk et al., 2010**), described in Section 2.

4.1.1. Issues Discussed

4.1.1.1. The Relevance of the Issues

The nurse supervisors engaged mostly with the issue of relevance of the problem formulation, responding to students with comments such as:

"I see that you really have brought up a central approach by choosing an issue such as interactions with people with cognitive impairment" (comment in LMS, NS6, S2012)

"Diabetes is about giving responsibility to the patient" (comment in LMS, NS1, S2010).

The nurse supervisors naturally related the issues presented to their relevance in the workplace. They wanted the students to be critical of practices and suggest what they could further focus on in their daily work. Relating this process to the steps in EBP, one can say that this corresponded to step '0'.

The nurse educators did not get very involved when students wanted advice on the relevance of the problem they had chosen or planned to choose. For example students stated: "We have decided on the theme of type 2 diabetes in nursing homes, with a focus on nutrition and interdisciplinary cooperation" or "We have not quite reached any concrete problem and welcome any tips about this" (comments in LMS, NS2, S2010). The educators left such questions to the nursing supervisors, who were best placed to supervise on such areas.

4.1.1.2. Clarification of an Issue

The librarian did not get involved with the relevance of the students' problems, only on the refinement of them, which corresponded to step 1 of the EBP practice. She got involved in terms of whether it was easy to find research-based information on an issue or not; coming with comments such as: "I think that it will be easier with the first topic/problem". The supervisor asked students to be more specific, for example, by recommending that they decide whether they would search for diabetes 1 or 2. The nurse educator engaged by adding: "Could this be further sharpened?" (comments in LMS, L1, NS2, NE4, S2010).

4.1.1.3. The Subject Terms and the Information Seeking Process

When it came to choosing subject terms and the information seeking process, it was clearly the librarian, who took the greatest initiative to get students to reflect on what terms and sources they would choose to do the information seeking (comments in LMS, L1, S2010, A2010, S2011(1), S2011(2)), but the nurse educators were also very active (comments in LMS, NE1, S2012). The nurse supervisor did not get involved here and these questions were left to the librarian and the nurse educators. Step 1 and 2 must be considered together in this case, since it was about how to sort a student's research problem into a PICOT format and to perform the search in a certain way. The librarian was able to provide feedback such as: "Think about what words you would choose to apply and what databases you want to use before the session (in the learning group) next week" (comments in LMS, by L1, L2, L3 in S2010–S2012). "I perceive that you are stuck, because of the term you have chosen" (comment in LMS, L1, S2010). The nurse educator supported by saying that it must be structured and systematic searching and that students must demonstrate the application process (comment in LMS, L1, S2012).

4.1.1.4. Literature Selection

Nurse educators engaged with the selection of literature and sources. They emphasised the importance of formalities, such as attaching a bibliography, also that the focus should be on how the research was important for practice. They reminded the students that they had to find recent research. This was commented several times (NE1-4, S2010–S2012).

4.1.1.5 How to integrate research findings into practice in the workplace

Discussion focused on whether the students' findings were valid, important, if they would help the patient and if they could be integrated into the workplace practices (step 3–4). In some cases, the students gave the department tips about how practices could be developed. For example, one student group gave tips on how the department could provide a better sensory garden for the patients and the nurses welcomed the recommendations (presentation of the assignment, S2011). Another student group tried to give recommendations to their department based on urinary tract infection research from 1992 and this caused reactions and discussions in the department (presentation of the assignment, S2010). The librarian and the nurse educator also had the opportunity to contribute in the discussions.

4.1.2. Communication Patterns

The way the supervision sequences evolved with regard to how the supervisors related to each other is described below.

4.1.2.1 Personal and uninhibited communication

The nature of the dialogue between supervisors and students was personal and uninhibited. To give a few examples: "I agree with X", "Have you [X] got any tips?", "I'm struggling to find…", "Is that how you think it should be set up?..", "Hey X, I've searched for research on Z pages and found some good tips. Is this a reliable source?", "I'm, going very slowly here…." (comments by students in LMS, S2010–S2012).

4.1.2.2. A Culture of Sharing

The LMS in itself was a tool for sharing, but the willingness to demonstrate a culture of sharing was also evident in the texts. The students described how they shared their knowledge of information seeking with the supervisors several times during the practical period (comments in LMS, S2010–S2012). The librarian took on the role of highlighting good examples of approaching a problem or searching for literature. One example was posting a link to a document in which one of the groups had documented the entire searching process (comments in LMS, L1, S2010). In this way the students also had access to all the feedback from the supervisors. Another example was when the librarian encouraged nurse educators to share their knowledge at the same time sharing some of her own expertise on information searching: "What do the nurse educators find important when looking for literature?" (comment in LMS, L1,S2011). The lecturer urged students to show what they had written in the LMS to the nursing supervisors and asked them to give feedback. The librarian shared her arguments about the benefits of searching for information in scientific databases: "I see that you wrote that you have searched on Google. Searching on Google can work when .... However, if you want an overview of research in an area, you should search in scientific databases … if you want a selection of Nordic research, you can search …" (comments in LMS, L1, S2010).

4.1.2.3. Towards a Common Vocabulary

The vocabulary in the guidance changed underway as supervisors adopted each other's perspectives and used words and concepts they had learned from others. All supervisors helped to recommend search terms or to emphasise that the search should be done in a systematic and not a random way (comments in LMS, S2012). It can be assumed that the nurse educator and nurse supervisors had learnt this from the librarian. There was a trend for the nurse educator to use the same concepts as the librarian or change to the vocabulary. The term 'subject issue' was modified to 'search term', for example. In contrast, the nursing supervisor was giving tips for search terms, while also giving views on the issue and recommended literature on where the measures are described. The librarian entered the nurse educator's and the nursing supervisor's subject areas and commented that she thought students should be encouraged to think more about the issues and their formulation for themselves (comments in LMS, S2012). Often the nurse educator and nursing supervisor agreed. They could supplement later with how restructuring the students problem could happen by referring to what students should be familiar with in nursing practice (Comments in LMS, NE2, S2011).

It seems that the supervisors had an agreed way to respond to the students. Although the librarians were not consistently thorough in how to do the supervising they usually avoided giving students a recipee, but rather asked the students to reflect. The nurse supervisor asked questions such as: "What do you think is most important to focus on?", "I would like you to point out things that are not working well and what's functioning OK?", "What do we need to focus further on in our daily work?", "Is it diabetes 1 or 2, diet, certain complications...?" (comments in LMS, NS2, S2010). The nurse educator asked questions and came with comments, such as: "Are you going to limit the topic?", "All three problems are about effect...". The librarian gave inputs such as: "I think you guys need to think a little bit more about the research question and rephrasing it", "I'm going to give feedback when you've worked a little further with the issue", "Now you have even come with a research question...", "I'm thinking that it will be easier with the first research question", "What do you want to focus on?", "That's a great topic with music to help!" (comments in LMS, L3, S2011), The nurse educator also supported the student's in choosing the topic of music (comments in LMS, NS4, NS5). Other comments from the librarian were: "I can't wait to see the problem framed", "Think of the words you want to choose to search with and which databases you want go through before our session next week"(comments in LMS, L1, S2010), "I notice that it is as if you're being blocked by the terms you have chosen", "I want the argumentation for the choice of databases" (commented several times in LMS, S2010–S2012). Although the librarian tried to encourage the students to reflect on subjects as the nurse educators did, she also gave the students, more or less, a recipe for how it could be done. For example: "You can start by translating your search words into English, check with Svemed +..." (comment in LMS, L2, A2011, L3, S2012). In this case, the students received the recipe, rather than having to find the way forward themselves and so receive supervision in the next session.

4.1.2.4. Acknowledgement of Each Other's Competences

Recognition of each other and other's expertise was mentioned several times. A nurse educator supported the librarian and that way gave her recognition as a partner: "I support X [the librarian], it's our task as educators to sensitise students in the work of collecting articles and the like. That's why we devote resources to education..." (comments in LMS, NE4, S2010).

Other examples of recognition of the librarian's inputs by the nurse educator included: "I agree that the problem is still too broad and that you can be more precise ...", "My suggestion is that you start searching through research together with the Learning Center", "Check the blog to see how you can get help from the Learning Centre", "Also read how to conduct clever searches in Wiki" (comments in LMS, S2012).

The librarian often referred to the nurse educator and made it clear that the nurse educator had more authority and knowledge to decide.

A nurse educator recommended that students benefited from expertise outside the community of practice, "I can recommend a researcher in nursing and ..... Her name is X and you will meet her during your nursing training. If you want, I can ask her if she would comment on questions from you via the blog", "The doctors at Y are also a good source ask" (comments in LMS, NE1/PM, S2010, S2011(1)).

4.1.2.5. Support to the Students

An important part of the repertoire of the trainers was student encouragement. All three groups of supervisors gave the students a lot of support and praise. Examples included highlighting that students had made good progress, a nurse educator writing that she is looking forward to the next step with a specific learning group and stating that something is "wonderful" or "looks promising". Other examples included encouraging words from the librarian such as: "That worked well!", "Great! You are very clever!". These comments appeared several times in LMS by all the supervisors, S2010–S2012.

4.2. Negotiation for a Joint Enterprise

This section describes the framework and measures that were present or were implemented that facilitated the librarian in becoming an integral partner in the supervision team.

4.2.1. Participation in the Whole Process from the Planning Phase, to Implementation, to Evaluation

What influenced the development of joint supervision practice in IL was communication in meetings about how to organise the practice periods. The head of the workplace, professional development nurses, nurse supervisors, nurse educators and the librarians all participated in the planning phase of the project. The participation took place in steering group meetings, reference group meetings and at meetings of the working group (for example meeting notes in January, October, November and December 2010).

One of the specific issues discussed and that all had mutual responsibility for was the design of a programme for each practice period (the programme for the practice period in LMS, S2010–S2012).

The schedule, types of supervision activities and responsibilities were defined collaboratively. The programme was uploaded onto the LMS, so that everyone could have access to an up-dated version of the document.

The project manager commented that: "Through trial and weekly exchange of experiences, adjustments and improvements appear to happen much faster and more easily" (Interview, PM, November 8, A2011).

4.2.2. Openness for Establishing an LMS for Better Continuity in the Supervision

The idea of adopting a LMS came from the librarian in the spring 2010, because the Learning Centre experienced the need to be available outside physical meetings (meeting notes, February 16, S2010). One of the arguments was also that every participant could support the supervision together in LMS as a supplement to supervision in physical meetings. The project manager of the Langerud model responded positively to the idea, before she knew what a blog was. She immediately saw the potential of the LMS for achieving greater continuity of supervision and having digital access and sharing of experiences in writing. Institutional Management was also immediately positive to trying out this new way of communicating, as a supplement to the supervision in the meetings of learning groups (presentation at the steering group meeting, February 16, S2010). This idea of providing supervision through the LMS turned out to require a greater effort from the nursing supervisors, since there were differing opinions on its purpose and because some would not share their expertise and experience there. Some were daunted about sharing on the LMS.

It was considered that the workplace should encourage their employees to participate in the LMS alongside their participation in direct supervision. Resistance was expressed by several nurse supervisors, because they did not think they had got enough information and training: "Just one of us has been in (the blog) and read a little, but didn't dare to write anything" (comment in LMS, May 31, S2010). Several courses in how to use the LMS were held by the librarian (for example courses, A2010, S2011). At the end of the project period in 2012, 15 of about 50 nursing supervisors had written in the LMS.

4.2.3. Seeing the Need for Greater Competence in IL and EBP

On the initiative of the the librarian and project manager for the Langerud model, an interdiciplinary group was established, which could focus on postgraduate education in EBP (note, June 8, S2010). This was easy to achieve, because it was agreed to by all of the participants. After the interdisciplinary group had completed the postgraduate education in the autumn 2010, a resource group in EBP was established (note, February 11, S2011(1)). This group also put in place new measures and fixed the days for classes in information searching at the beginning of each practice period (for example, a meeting note from January 2012). These courses were primarily arranged for the nurse supervisors, but nurse educators and students were also permitted to participate (course, S2011, S2012).

4.2.4. Agreement and Disagreement about the Aim to do the Information Seeking in a Certain Way

Most agreed that the aim of the supervision should be that the students use recent research for their assignments. However, not everyone agreed on how students should conduct this (meeting notes and e-mail, A2010). Not all agreed with the proposal to use the seven steps to apply a system and improve the quality of outputs. The librarian had to spend some time generating understanding in the project group that the quality of search results is dependent on taking a systematic approach to searching in scientific databases, as opposed to doing random searches in the same databases, or doing searches in Google. The manager of the Langerud model stressed that the requirement had been that students should do systematic information seeking and use research articles for their assignments, throughout the project period (guideline in IL instruction in for the supervisors, LMS, S2010). Despite this, there were some, who did not see the importance of it for first year Bachelor's degree students. One nurse educator thought it was sufficient to search on websites for health services and on Google. In this case, the librarian justified the importance of having a relevant amount of terms and doing the search on scientific databases.

The nurse educator had greater authority than the librarian and gave her own guidance about how the students should conduct searches. In one learning group meeting, the importance of the nurse educator's authority was emphasised and, as a result, the students did not deliver assignments based on up-to-date research (S2010). Consensus was, however, achieved on this student requirement in the next training period (meeting notes, S2011). A reason for this could be that the participants had become more familiar with each other's competences. They had also learned a lot about informaton searching during the implementation of joint courses and interdisciplinary supervision.

4.2.5. Agreement on Supervision for Reflective Thinking

Students were trained to reflect about: 'why we do what we do?' and come with positive criticism to the practice (guideline in IL instruction for the supervisors, LMS, S2010-S2012). In this way, they developed their interest and commitment to find up-to-date research to improve practice, or confirm that current practices were correct. Students were required to identify issues and follow the seven steps of evidence-based practice to examine them. The institutional head of the work placements, the project manager for the Langerud model and the librarian were all present at most of the introduction sessions. From day one, the students were encouraged to start reflective thinking in the*Langerud model.* The training format was brought into meetings with the learning group and the students. In this way, the assignment was linked more closely to practice and tied to student's clinical experiences.

4.3. Mutual Engagement

Mutual engagement did affect the joint supervision and the findings are presented below.

4.4.1. Diverse Backgrounds, Experiences and Roles

Opinions of participants differed as to what makes a community of practice possible and productive. Diverse backgrounds and experiences did allow one not only to use one's own skills, but also to appreciate the expertise of others. This was expressed especially by the project manager, who was also a nurse educator as presented below (interview, PM, November 8, A2011).

Once participants have more equal roles, contributions can overlap. This was the case in this study. For example, the project manager and nurse educator commented: "as far as I see it, collaboration is important both in the project group meetings and in meetings with the learning groups. Developing a programme together with different opinions, strengthens the practice approach ... one can share experiences, which can help nurse educators become more coherent". Co-teaching, developing something together and sharing experiences were important to generate interest. The project manager stated: "It's important to create learning on the basis of the situation one is in. I'm not the one, who has the solution, we have the solution together". The librarian tailored supervision based on what the others commented. One nurse educator stated that: "Competence is greatly expanded by having several professions. Something happens across them". The librarian pointed out that there must be mutual transparency for collaboration to function. She thought that everyone should have the opportunity to contribute supervision in their own way and at the correct moment in the training. "Students sat up when you [the librarian] spoke", one of the nurse educators noted (observation notes, S2010). She expanded saying: "professionalism is contagious...we can get wider and deeper expertise together".

The reflection of the nurse educator, who was also the project manager and collaborated most with the librarian, implies that this happened: "What made it strong was synergy, it's double, we see it from different angles. Then we all get dragged in. It became the motivating force". The nursing educator led the meetings with the learning groups, but she supervised together with the nursing supervisor and the librarian. The librarian did not have a specific time slot in the two learning groups, she specifically engaged with, but attended all the meetings with these learning groups. The librarian was more engaged by being able to participate and input as an equal with the nurse educator, rather than having an allotted time for providing input. With a time limited input, the librarian would only contribute a portion of relevant expertise and would need to choose what to emphasise. In the integrated teaching format, she had a chance to engage in the supervision when it suited best throughout the entire training session. In this way, the different supervisors were able to complement each other (observation notes, S2010-S2012).

4.4.2. Becoming a Valued Member

One of the nurse educators pointed out that one became a part of the community through trial and weekly experience. The project manager of the Langerud model sometimes said to the librarian: "either you're present, or you're not" (Interview, PM, November 8, A2011). The librarian's experience was that it was important to be present to participate in communication with the others, otherwise you could miss something and did not capture what lay 'between the lines', the unspoken. In addition to participating in various weekly experience exchange meetings, it was also possible to keep up-to-date by reading the students' work, tutoring sessions and also provide supervision inputs through the LMS. As the project manager of the Langerud model said: "this is where it's happening". The LMS was used as a common place to share information. It actually generated such a degree of engagement that the head of the workplace wanted the LMS to keep running also after the project period ended (note, October 11, S2012). She wanted the librarian to continue to have responsibility for the LMS and supervise through it.

4.4.3. Relationships and Mutual Commitment

Mutual engagement creates relationships between people and mutual commitment to what is going to be produced. Mutual commitment is reflected in doing things together and this in turn maintains a community. The Langerud project manager articulated this: "We [all the participants in the study] together created a need to do information seeking" (Interview, PM, November 8, A2011). It was felt that the students and nurse supervisors would see the same need during the course of the project. The commitment created by building on students' own experiences of practice was also seen as vital for contributing to the supervisors' motivation. One of the factors which generated respect was the attitude of the nurse educators to students. The project manager verbalised this approach: "Every contribution from students must be met with respect". Students were encouraged to be critical and to reflect on practices. They could debate why things worked as they did. Taking this approach the supervisors developed and reviewed the course with the students. One of the nurse educators stated: "We are helping to support the departments' improved quality. We are part of something important" (Interview, PM, November 8, A2011). In several meetings, it emerged that nurse educators may also have a role in encouraging nursing supervisors to use the LMS as a tool. It was evident, through inputs in the LMS training sessions that the nurse educators were already committed to using the LMS. This was in contrast to the lower participation of the nurse supervisors (see Table 1). Nurse educators commented a total of 85 times, whereas nursing supervisors only contributed 26 times. The librarian dominated with

115 comments. This was perhaps natural, since most of the questions were about the information seeking process. To use LMS was an initiative of the librarian for trying cross-collaborative supervision, so maybe the librarian felt a strong responsibility to make it function.

Table 1. Who communicates in the LMSs? (LMSs, S2010–S2012).

Who communicates?	Number of comments
Students	59
Nurse educators	85
Librarians	115
Nurse supervisors	26
Total	285

The supervisors were motivated by having support from other supervisors. Commitment to getting more people to supervise through the LMS was motivated by one of the nurse educators: "Do the nurse supervisors have any comments on this issue?" (comments in LMS, NE1, S2011). The librarian was also endorsed by a nurse educator: "I support the librarian, it's our responsibility as educators to sensitise students in their work when they do the information seeking" (comments in LMS, NE4, S2010). This illustrated that there was a commitment to teach students to find research, but also an acknowlegement of the librarian's expertise. The supportive expressions came several times during the project period, so these are just a few examples.

4.4.4. New Identity

The project manager of the Langerud model stressed the importance of several professions collaborating and supplying complementary and overlapping competences. Each individual provided expertise and knowledge, which were expanded by the expertise and knowledge of others. In this way, each member reshaped his or her identity in the new community. The librarian's contribution was summed up by the manager: "You have helped to build up an expertise that means that we now have a totally different way of talking about searching, you have helped to build up competence that also enables us to integrate this into our work" (Interview, PM, November 8, S2012). This statement also conveys something about the changed role of the librarian in the community.

5. DISCUSSION

Until this project started, the librarians at Oslo University College3 had only taught students on campus. By participating in collaborative supervision during the students' work placement, the librarian's role was changed and challenged by several factors. The findings are discussed with a focus on how a librarian can become a teaching partner during work placements.

5.1. Communicative Interaction among Supervisors

The co-supervisors had three different vantage points; education, the workplace and the library; each with their own operating practices. The professional identity of individual practitioners lies behind and manifests itself in current practice (**Gleerup, 2011**, p. 14). It involves frames placed around the professional practices, such as values, patterns, roles, communication messages and attitudes. The concept emphasises the potential of the work to develop relationships and continuity between profession and education (**Heggen & Damsgaard, 2010**, p. 86). Professional identity is also extensively shaped by habits and truisms and becomes part of what we do without being so conscious of why. Participants found their place and acquired a new identity. Identities joined and connected via mutual engagement for teaching content in contexts related to students' perspectives. The importance of a close relationship between content and context was emphasised by**Limberg and Sundin (2006)**. When the three groups of professionals met, there were often discussions and questions about the 'why', because they each had different angles of interpretation. Another reason could be that the meetings did not take the format each party was used to, but focused on the, so far, unknown or unformulated issues in the context of the inter-institutional or cross-professional initiative. Suddenly, new frames and new ways of speaking developed; one could see and touch each other's perspectives. The supervisors had to release their own 'language room' to get to know a new language and become part of a new space. In the new space, professions met with their assumptions, attitudes, expertise and ways of asking. The diverse supervisors approached each other in a new way of talking together. Gradually, they created a common repertoire that made them both understand each other better and guide the students collaboratively. The way in which the supervisors engaged with each other's fields was evident in the written inputs on the LMS. It was useful to have both different and similar roles in a community of practice to synergise effects in the supervision (**Wenger, 1998**, pp. 93–94). There was also a natural division of responsibility linked to who answered the students' questions. Supervisors developed a unified way of talking to encourage greater reflection from the students. The dialogues showed recognition and respect for each other and each other's contributions. Statements such as: "I'm not the one who has the solution, we have the solution together", showed that there was a development in the attitudes about how to learn together and share knowledge (interview, PM, November 8, A2011). These experiences can be seen as positive, but they also indicate that the supervisors had learnt to develop solutions together. This is consistent with Wenger's statement that diversity allows us to see other's expertise. Such experience was also confirmed in the study at Sonoma State University, where cooperation with an IL specialist was described as: "synergic—producing end findings that are more than the sum of the teaching-parts" (**Brodsky & Bower, 2005**). The Ball State Nursing Programme also demonstrated similar findings with the valuing of complementary expertise and sharing of knowledge (**Florea, 2008**).

The fact that the librarian had the opportunity to participate in the entire process, from the planning stage to the implementation of supervision and evaluation, was influential in her becoming a valued member of the teaching team. This process contrasted greatly to how the collaboration with the faculty worked on campus, when the librarian only collaborated in the planning phase and taught alone (**Mounce, 2010**). **Wenger (1998)**states that to become a valued member of a community of practice, it is not enough to know and understand the latest professional work. One also needs to become part of the social community. The importance of being a part of a process is pointed out in previous research. **Lloyd (2013)** emphasised co-participation in the shaping, negotiation and renegotiation of the community as crucial. **Shank et al. (2011)** highlighted the importance of being an integral, educational partner and in UCLA's Freshman Clusters, the librarian collaborated with the faculty on everything; from content in web sites to supervising in discussion for a (**Lindholm, 2007**). Through participating in the whole process, the librarian acquired increased insight into how she could contribute with her own competences and customise them to other's expertise. She found that there was a need for competence raising measures and new methods for better interaction; such as using LMS for learning and communication.

The librarian's proposals for information training courses and the establishment of the LMS were accepted and implemented. This was needed in order to shape a better starting point for the supervision of students and have the ability to supervise the students throughout their practice period. The communication patterns and the common vocabulary that emerged were made possible by the common supervision platform. The LMS strengthened the learning and communication. The proposal to set up an LMS was approved despite the fact that several participants did not initially know what a 'blog' was. The findings from this study and from Florea's research (2008) show that collaboration through LMS strengthened the collective responsibility for students' learning. By the end of the study, the institution leader saw this as indispensable. The Langerud model project manager's statement that the librarian had helped develop an expertise that led to a new way of talking about information seeking and integrating it into practice, confirmed that the librarian had been integrated, accepted and was respected as part of the joint supervision community. Confidence is also emphasised as one of several important prerequisites for collaborative teaching partnerships (**Ivey, 2003**). The fact that the librarian was shown great trust was a reflection of the confidence developed amongst co-supervisors.

5.2. Towards Co-teaching

The librarian's new collaborative constellations with faculty are described in several ways in recent research. Terms such as 'embedded librarianship', 'embedded library instruction' and co-teaching have been discussed (**Tilley, 2011**; **Vassilakaki & Moniarou-Papaconstantinou, 2015**).

Tilley introduces the term co-agency as a prerequisite to increase the good quality of co-teaching. Co-agency is about developing good relationships in a learning community. Co-teaching is all about bringing together different expertise and developing a product for the benefit of the target group. Co-teaching differs from the more traditional way of doing teaching, which involves faculty and library staff planning lessons and content in advance of the librarian's teaching, but not teaching together.

In this study, as described above, a common repertoire in the supervision setting was developed. Wenger mentions the importance of enhancing already existing knowledge, so that colleagues can be acknowledged and their skills can be used in interacting with others.

These skills can often be invisible for the colleagues otherwise. This was highlighted several times between the supervisors, both in meetings and through the LMS. New knowledge was created by utilising the distributed knowledge. **Wenger (1998)**sees difference as a resource for creating mutual commitment. This can be particularly apparent when participants have different roles, backgrounds and knowledge. The reciprocal commitment can then lead to complementary contributions. The support of students, while they are doing clinically oriented information seeking and academic oriented information seeking (**Sundin, 2003**, pp. 173–175), can promote collaboration between education and the workplace.

The findings indicate that the structured way of doing supervision both in meetings and through the LMS strengthened the 'sharing culture' and the collective responsibility for students' learning. This is in line with Wenger's claim that adapted structures, in this example the LMS, will only be effective if participants integrate them into their communities.

A natural sharing of themes among the supervisors developed. All the supervisors had one common place to find all the information and keep up-to-date themselves. The LMS can particularly support students in the process of becoming information resilient workers (**Florea, 2008**; **Tuominen et al., 2005**).

The positive personal relationship between the researcher/participant observer and informants may have influenced the positive results. The researcher may empathise with her informants and vice versa. It was easy to gain support for ideas and implement them.

6. CONCLUSION

If the librarian is to be a partner and a catalyst of IL workplace learning, together with nurse educators and nurse supervisors, there is a need to establish both interdisciplinary co-agency and co-teaching. The combined willingness and ability to collaborate and use complementary competencies to provide better quality instruction was a key to success. The mediating tools had an important role in this process. The LMS stimulated joint supervision culture by being a 'meeting place' for learning and communication. The need for improved capacities in IL and EBP was addressed in this study. However, if librarians are to engage with interdisciplinary supervision on an ongoing basis, it is relevant to ask whether they need further pedagogical training. Alongside this, there may be a need to focus on technological tools that are suitable for supervising and that can support purposive pedagogical and didactic approaches.

NOTES

[1]Changed name in 2011: Oslo and Akershus University College of Applied Sciences.

[2]Changed name in 2011: The Learning Centre and Library.

ACKNOWLEDGEMENTS

I would especially like to thank Gerd Melsæter, the project leader for the Langerud model, who invited me to join the project and was jointly responsible for the results. In addition, I am very grateful to the participants of the Langerud project for enabling me to do this research as part of their group. I would also like to thank the Director of Learning Centre and Library, Lars Egeland, for encouraging this project to happen. This text would not have been realised without the help of my critical readers Professor Ragnar Nordlie and Senior Professor Louise Limberg. Advisor Nicola Johnston has also been indispensable along with the anonymous reviewers, who have given constructive feedback to further elaborate and improve the accessibility of this article.

REFERENCES

1. American Psychological Association. (2010). Publication manual of the American Psychological Association. Retrieved June 16, 2015,

2. Association of College & Research Libraries. (2000). Information literacy competency standards for higher education. Retrieved September 17, 2015,

3. Association of College & Research Libraries. (2013). Information literacy competency standards for nursing. Retrieved September 17, 2015,

4. Bonner, A., & Lloyd, A. (2011). What information counts at the moment of practice? Information practices of renal nurses. Journal of Advanced Nursing, 67(6), 1213–1221.

5. Brasley, S.S. (2008). Effective librarian and discipline faculty collaboration models for integrating information literacy into the fabric of an academic institution. New Directions for Teaching & Learning, 2000(114), 71–88.

6. Brodsky, K., & Bower, E. (2005). Information literacy: What are we really talking about? (Crux brochure for FYE): Sonoma State University.

7. Cheeseman, S. (2013). Information literacy: Foundation for evidence-based practice. Neonatal Network, 32(2), 127–131. doi: 10.1891/0730-0832.32.2.127.

8. P., Cartabellotta, A., Martin, J., Hopayian, K., ... Osborne, J. (2005). Sicily statement on evidence-based practice. BMC Medical Education, 5(1), 1. Retrieved September 21, 2015,

9. Florea, M. (2008). Using WebCT, Wiki Spaces, and ePortfolios for teaching and building information literacy skills. Journal of Library Administration 48(3/4), 411–430.

10. Forster, M. (2013). A phenomenographic investigation into information literacy in nursing practice — Preliminary findings and methodological issues. Nurse Education Today, 33(10), 1237–1241.

11. Gleerup, J. (2011). Viden(skabs)teori. In N. B. Hansen & J. Gleerup (Eds.), Videnteori, professionsuddannelse og professionsforskning (pp. 83–118). Odense: Syddansk universitetsforlag.

12. Heggen, K., & Damsgaard, H. L. (2010). Kvalifisering for profesjonsutøving: sjukepleiar—lærar—sosialarbeidar. Oslo: Abstrakt forlag.

13. Ivey, R. (2003). Information literacy: How do librarians and academics work in partnership to deliver effective learning programs? Australian Academic & Research Libraries, 34(2), 100–113. Retrieved September 21, 2015,

14. Kolstad, A. (2009). Flexibility in teaching information literacy by using wikis. Paper presented at the IFLA satellite meeting, Helsinki. Retrieved June 16, 2015,

15. Limberg, L., & Sundin, O. (2006). Teaching information seeking: relating information literacy education to theories of information behaviour. Information Research, 12(1), 7. Retrieved September 21, 2015,

16. Lindholm, J. (2007). Four years later: reflections on freshman cluster experiences. Los Angeles: UCLA Office of Undergraduate Evaluation and Research, Division of Undergraduate Education. Retrieved September 21, 2015,

17. Lloyd, A. (2013). Building information resilient workers: The critical ground of workplace information literacy. What have we learnt? In S. Kurbanoğlu, E. Grassian, D. Mizrachi, R. Catts, & S. Špiranec (Eds.), Worldwide commonalities and challenges in information literacy research and practice (Vol. 397, pp. 219–228). Cham, Switzerland: Springer International Publishing.

18. Melnyk, B. M., Fineout-Overholt, E., Stillwell, S. B., & Williamson, K. M. (2010). Evidence-Based Practice: Step by Step: The Seven Steps of Evidence-Based Practice. AJN The American Journal of Nursing, 110(1), 51–53.

19. Mounce, M. (2010). Working together: Academic librarians and faculty collaborating to improve students' information literacy skills: A literature review 2000–2009. Reference Librarian, 51(4), 300–320.

20. Rapley, T. (2007). Doing conversation, discourse and document analysis. London, SAGE.

21. Sackett, D.L., Rosenberg, W.M.C., Gray, J.A., Haynes, R.B., & Richardson, W.S. (1996). Evidence based medicine: what it is and what it isn't. British Medical Journal, 312, 71–72..

22. Säljö, R. (2005). Lärande och kulturella redskap: Om lärprocesser och det kollektiva minnet [Learning and cultural tools: on learning processes and collective memory]. Stockholm: Norstedts akademiska.

23. Shank, J. D., Bell, S., & Zabel, D. (2011). Blended librarianship: [Re]envisioning the role of librarian as educator in the digital information age. Reference & User Services Quarterly, 51(2), 105–110.

24. Sundin, O. (2003). Informationsstrategier och yrkesidentiteter: en studie av sjuksköterskors relation till fackinformation vid arbetsplatsen (Vol. 25). Göteborg: Distribution: VALFRID.

25. Tilley, E. (2011). Community learning: the story of adopting co-agency for effective learning in a UK university faculty library. New Review of Academic Librarianship, 17(1), 129–148.

26. Tuominen, K., Savolainen, R., & Talja, S. (2005). Information Literacy as a Sociotechnical Practice. Library Quarterly, 75(3), 329–345.

27. Vassilakaki, E., & Moniarou-Papaconstantinou, V. (2015). A systematic literature review informing library and information professionals' emerging roles. New Library World, 116(1/2), 37–66.

28. Wenger, E. (1998). Communities of practice: learning, meaning, and identity. Cambridge: Cambridge University Press.

10

CHAPTER

Digital Scholarly Publishing and Archiving Services by Academic Libraries: Case Study of the University of Patras

Panos Georgiou, Giannis Tsakonas

Library & Information Centre, University of Patras, Greece

ABSTRACT

During the last years, dramatic changes in the electronic publishing landscape have created new roles and changed the traditional ones. Presently, some libraries have capitalised on their experience and knowledge in information technology and electronic publishing to undertake such activities, while at the same time they spearhead the campaign for Open Access spreading within academic communities. The Library & Information Centre (LIC) of the University of Patras (UoP), Greece, has been playing an active role in promoting Open Access (OA) in Greece. Since 2007, LIC has been experimenting with OA publishing practices and tools within the framework of various R&D projects. Two of the major results of these efforts are the 'Pasithee' e-publishing platform and the 'Dexamene' digital archive for Greek scholarly journals. Both platforms are based on OJS-Open Journal Systems e-publishing software. The two facilities were appropriately modified to meet the LIC's publishing and archiving requirements respectively. Currently two journals are being hosted on each platform and all four are from the Humanities. The LIC is negotiating with more publishers and editorial teams to host their journals. In this article we focus on: - technical and managerial key issues of the development and operation phases, - services and procedures, - the business model, - technological, procedural and legal issues and problems that were encountered when working together with publishers, editors and authors, and - future plans for improving and upgrading our e-publishing services into an integrated institutional platform to cover all kinds of publications and data types (monographs, conference proceedings, teaching material, bulletins, magazines etc.). The article concludes with a succinct presentation of the Directory of Greek Digital Resources, a pilot infrastructure developed by the LIC which indexes and presents digital publishing initiatives in Greece and aims to become a formal registry of Greek scholarly resources in digital format.

Keywords: electronic publishing, academic libraries, Greece, Open Access

In this article we focus on:

1. technical and managerial key issues of the development and operation phases,
2. services and procedures,
3. the business model,
4. technological, procedural and legal issues and problems that were encountered when working together with publishers, editors and authors, and
5. future plans for improving and upgrading our e-publishing services into an integrated institutional platform to cover all

kinds of publications and data types (monographs, conference proceedings, teaching material, bulletins, magazines etc.).

The article concludes with a succinct presentation of the Directory of Greek Digital Resources, a pilot infrastructure developed by the LIC which indexes and presents digital publishing initiatives in Greece and aims to become a formal registry of Greek scholarly resources in digital format.

INTRODUCTION

Libraries were always considered to be the custodians of knowledge by preserving and giving access to the media of each age, either papyri, codices, or books. The last thirty years of electronic and digital technology's rapid development have created new roles and challenges for libraries. As organisations that rely heavily on the 'technology' of each age, the social conditions and the economic parameters, libraries were always trying to find new means to better serve their users. This is where libraries' motivation for transformation originates, as they are organisations that work closely with their community and therefore many of the libraries' achievements are based on their needs. Libraries have often been criticised in the past for taking part in initiatives such as developing institutional repositories or participating in institutional e-learning programmes. However, they have managed to deal with this 'identity crisis' and their activity in these fields proved fruitful, mainly due to the perception of library managers that these new roles are an 'expansion rather than contraction' (Jöttkandt, 2010).

Nowadays the library's identity crisis takes place in a wider environment of crisis. These last two years the economic downturn is piling on the pressure of instability and budget shrinking. Libraries see their budgets decreasing and suffer from further reductions each year, while at the same time the amount of scholarly information increases and more requests to expand the collections are being submitted and considered. It is widely known that journal budgets represent the lion's share of the collection development budget. According to a report of LISU and Loughborough University, journal prices in biomedical sciences increased from 42% to 104% and in the social sciences from 47% to 140% (White & Creaser, 2007). The study investigated the increase per journal title, per page, percentage of Impact Factor coverage and per point in the Impact Factor during the years 2000–2006. A recent statistics study by ARL (2009) notes that the expenditures among 110 members of the Association have increased from $49,497,141 in 1997–1998 to $554,637,844 in 2007–2008. While the numbers are subject to yearly variations, we still see an increase of 1.020%. The recent library economics landscape shows that despite 'freezes' or slight variations in journal prices for 2010 (Henderson & Bosch, 2010), libraries still face significant problems due to their own overall budget reductions and the new titles entering the subscriptions 'arena'.

The present environment, however, does not only present challenges to libraries, it also creates new opportunities which are based on technologies and related interventions. Technological advancements and emerging political issues have helped libraries to reinforce their relationships with scholars and develop services that deal with the entire information lifecycle rather than just the end. Many fields of assistance with regard to creating, disseminating and using information are still open and recent studies have shown that librarians are eager to explore new roles for their profession and the service of their community (Brown, Griffiths, Rascoff & Guthrie, 2007). On the other side, scholars, who are an integral part of the information lifecycle, need support from a reliable partner that is willing to share know-how and adaptable services. Thus, through long-term cooperation and service, libraries have developed a keen understanding of the interests of scholars and their need to ensure the most effective and widest impact of their research. Scholars understand that the current publication models are limiting their readership, but despite recent evidence to the contrary, they still fear that Open Access publishing will negatively affect the impact of their work and therefore they seem reluctant to move to OA. However, a growing number of scholars are becoming aware of both the benefits and drawbacks of Open Access and are seriously considering their options (SAGE, 2009).

Several libraries are taking on electronic publishing in response to the overall budget crisis and the need to satisfy specific classes' requirements. And the stage for such endeavours is set when one takes into account the library budget reductions, the journal prices increase and the available alternatives; all that is needed is a motive. The present article summarises the recently inaugurated e-publishing programme of the Library and Information Centre of the University of Patras (LIC). The following section first describes important case studies from the library domain that work in an exemplary fashion and then outlines the Greek academic publishing landscape. Section 3 presents the case of the LIC by listing the conditions in which the e-publishing programme takes place, the activities carried out, the implementation details and the outputs. The last section is dedicated to discussing the effectiveness of the e-publishing programme and the future challenges.

BACKGROUND

Libraries as Publishers

There are several examples of publishing activity by library organisations. Many of these initiatives are based on partnerships between libraries and information technology centres and/or university presses. The organisational schemes in which they operate are often idiosyncratic. There are instances in which the publishing initiative is subsumed in another organisational scheme, like the general publishing vehicle of a university, which may include institutional repository activities, like in the case of the Australian

UTSeScholarship,[1] or even to the library itself, such as in the example of Linköping University Electronic Press.[2]

Project Euclid,[3] funded by the Andrew W. Mellon Foundation and operated by Cornell University Library and Duke University Press, is a digital library of electronic journals in the disciplines of mathematics and statistics. According to the project's webpage 'Project Euclid is designed to address the unique needs of low-cost independent and society journals', thus seeking a balance between efficient economic management, scientific integrity and wide publication. Project Euclid is based on the DPubS (Digital Publishing System) platform, an open-source software system. DPubS was developed on the Dienst system, which was the responsibility of Cornell's Computer Science Department and Cornell University Library, while in its current form DPubS is supported by Penn State University Libraries and Press. Project Euclid represents an example of reasonable responsibility sharing between peer partners for the effective publishing of high-quality journals, while respecting policies of individual societies.

The Scholarly Publishing Office of the University of Michigan Library is an interdisciplinary publishing service that focuses mainly on the humanities and social sciences.[4] One of the most important journals in the scholarly communication field is published under the auspices of the SPO, namely the Journal of Electronic Publishing, while other publications of the programme range from encyclopedias to archival material. The office provides (a) hosting services for non-profit academic organisations that are interested in publishing, (b) infrastructure for an online delivery system that secures accessibility and retrieval, and (c) a selection of statistical indicators. According to the official statement SPO is seeking 'to foster a better economic model for campus publishing, to support local control of intellectual assets, and to create highly functional scholarly resources'. The publishing programme is also framed by a wide set of services to support alternative dissemination channels, such as print-on-demand services for selected items from the collections of the University of Michigan digital library. Once again the Office provides the option to interested parties to follow several access policies, like open access or pay-per-view for non-subscribers.

SPO has extended its publishing model by partnering with other initiatives to promote Open Access. Open Humanities Press,[5] an international initiative that promotes electronic publishing for journals and books, makes use of the infrastructure and experience of SPO and attempts to stimulate scholars, editorial teams and research societies to participate in a venture for facilitating editorial work. The approach is fully Open Access with no limitations or charges.

These examples constitute long-established and solid endeavours with different business plans. For instance, Project Euclid and SPO grant tiered access rights, ranging from fully open to subscription-based access, while OHP is providing its services free of any charge, either to the authors, or to the users. Therefore the publishing policies vary according to contextual parameters that are often based on the long-term traditions of the institutions and the scholarly societies.

Scholarly E-Publishing in Greece

Scholarly publishing in Greece is mainly concerned with the printed format. According to data from the National Book Centre of Greece (EKEBI) (EKEBI, 2009), the Greek book market is still a completely print environment and is mainly oriented to the humanities and literature, as titles in natural and applied sciences represent a small proportion of the total production and even seem to be declining in numbers. 328 journals have been recorded (Georgiou & Papadatou, 2009) as scholarly titles or titles of a considerable scholarly interest and importance to academics and researchers, based on criteria such as the content, the editorial team, the publishing body, the subject, the review policies, the publishing procedures and in some cases the history of the journal. Only 163 of these titles adopt a clearly stated peer review policy in the article submission, editing and publishing processes.

According to the same report, there has been a significant progress regarding e-journals publishing initiatives in the last two years, while activities in the e-book market are considered rather insignificant, although small increasing trends can be observed.

Out of a total of 118 journal titles with online full text, 84 are providing their contents through OA, and the vast majority of the latter (73 titles) are explicitly identified as peer-reviewed journals (Figure 1).

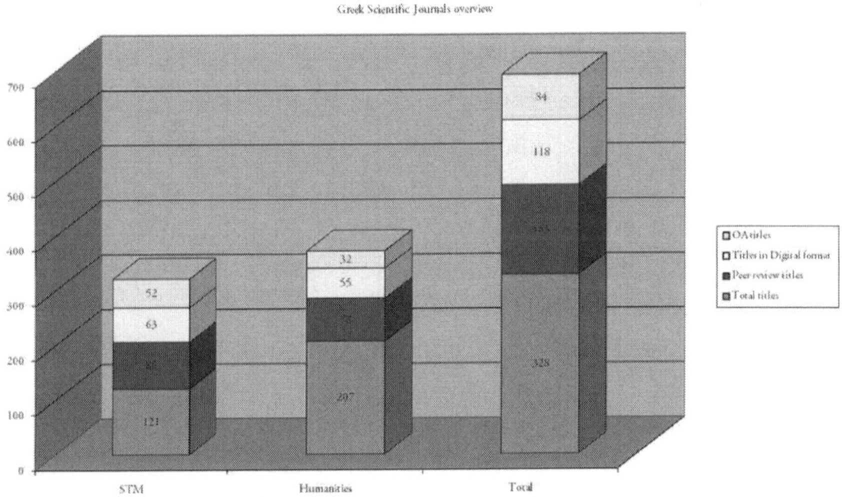

Figure 1. Overview of Greek scholarly journals.

Almost 40% of these 118 journal titles are published both in print and online, while 36% are practically digital archives of active journal titles. Fully electronic journals represent only 24% (29 journals) of the total number of digital titles of which all but one are OA journals. The majority (~40%) of these journals are published in the field of medicine (mainly by related learned societies), with journals in the area of natural and applied sciences accounting for a small number of titles.

Overall, the various societies seem to be more active in digital publishing than other types of organisations and they lead the race in electronic-only journals along with the academic institutions. Private publishing houses hold a significant share of the total online market (23%), although they are clearly oriented to 'print & web' publications.

The Case of the Library and Information Centre, University of Patras

The current section presents the case of the Library and Information Centre (LIC) of the University of Patras. The LIC is one of the pioneer libraries in Greece in the adoption of OA, taking actions itself or coordinating the institutional planning for OA. This OA policy is reflected in numerous ways: the LIC is a Berlin Declaration signatory, a SCOAP3 Initiative supporter and a BIOMED Central Supporter. During the last years the LIC has cooperated with several organisations to produce high-quality digital collections, a.o. schools and departments of the University of Patras, local information organisations like the Municipal Library of Patras, non-governmental agencies such as the Hellenic Literature and History Archive, and governmental agencies like the Library of the Greek Parliament. These collections are outlined in the following subsections.

Context of Activity

In the last decade the LIC developed a series of initiatives in the fields of e-archiving, e-publishing and scholarly communication. Most of the collections are deliverables of the EU co-funded R&D project 'Telephaessa'. 'Telephaessa' ran over a six-year period for the first phase (2000–2006) and was extended to 2007–2008 in order to do further work on some successful activities. Among others, the objectives of 'Telephaessa' were: 'the acquisition of technical expertise and infrastructure on issues of digitisation and digital service delivery to (LIC's) users' and 'active participation in initiatives for the enrichment of digital collections in the Greek language on national and international level'.

It was firmly decided by the LIC to develop the collections and services with an 'open' perspective and to encourage participation in cooperation schemes with other local and regional state services. Aspects of 'openness' were considered to be (a) the prioritisation of open source software, (b) the pursuit of means to effectively facilitate widespread distribution of open source at a national level, (c) the adoption and reuse of 'open' constructs, and (d) the adoption and dissemination of OA policies.

The R&D project of the LIC resulted in a set of collections that can be classified in three main categories according to their aims and scope:

- E-Archiving collections that host, preserve and disseminate old literature periodicals. Three collections are included in this category, namely 'Kosmopolis',[6] 'Pleias'[7]and 'Danielis'.[8] 'Danielis' is an example of the desired outcome regarding participation in local and regional cooperation schemes, as it was developed in cooperation with the Municipal Library of Patras and aimed to cover a significant gap in the field of digital resources for local history and culture by means of the digitisation of journals that were published in Patras during the 19th and 20th centuries.

- E-Publishing collections that support the publication of scholarly journals through digital channels. Two collections are included in this category, namely 'Dexamene'[9]and 'Pasithee'[10] (which are also the object of section 3.3).

- Scholarly communication and support services that host, preserve and disseminate the scholarly and research outcome through the employment of UoPatras' institutional repository

'Nemertes'[11] and UoPatras' Current Research Information System (the latter is still in a preparatory stage).

Figure 2 schematically presents the outputs of the 'Telephaessa' project. This figure also presents the activities undertaken by the LIC on the communal level, meaning the sum of all actions that support the dissemination of systems and collections (including translation services for the Open Journal System and Open Conference System in Greek, support for these systems' implementation, and installation and operation of the Wiki for the Greek DSpace community) and gather essential information about the current state of digital initiatives in Greece through the Directory of Greek Digital Resources (a brief presentation of the latter follows).[12]

Figure 2. Activities of the LIC towards Open Access at institutional and communal levels.

The activities of the LIC were encouraged by the lack of a proper digitisation centre and a publishing house within the University. To conclude, the motivation for the LIC to step on the e-publishing stage was not based only on the need to respond to the wide financial crisis, as was highlighted in the introduction of this article, but also on the need to assist scholarly communities in finding reliable and effective alternatives to publish in a world of emerging technologies.

IMPLEMENTATION DETAILS

Business Plan

The aim of the LIC's e-publishing programme was primarily to support academics or scholarly groups within the University of Patras to publish their research outputs through scholarly journals, but later it was extended to also address demands for storage and preservation of back-files of currently printed scholarly journals.

The LIC does not adopt an 'aggressive' policy to address scholars' needs, but rather works by encouraging a 'volunteer collective' approach (Jöttkandt, Willinsky & Kimball, 2009). Despite thorough work on recording scholarly activity, there are still margins where scholarly communication cells reside. Thus the LIC has issued an open call to interested parties to discuss possibilities for e-publishing, highlighting the benefits of lifting many cost-related weights and undertaking the responsibility for secure and reliable access. The call is open in many aspects, as there are — almost — no strings attached to the policy behind it. The only prerequisites are (a) an OA mode of publication and (b) safeguarding scientific integrity by following proper editorial processes and including peer review.

The LIC also offers a range of services that — despite their limitations — have been characterised as appealing, effective and viable by Greek standards. Due to limitations in resources and expertise, the business plan predicts some things that the library can do and some things that it can't. Among the tasks that the LIC can be responsible for is: (a) to digitise the material and import metadata, (b) to provide and guarantee access, (c) to provide infrastructures and software solutions, (d) to preserve the content, and (e) to instruct editorial boards on the administration of the system and the facilitation of the processes. The LIC does not at any stage interfere with the reviewing process, copy editing, parallel printing, distribution and subscription management, where applicable and only for printed versions. Finally, the LIC can provide consultation and assistance in developing a marketing strategy for promoting the journal and widening its contribution base, but this option is subject to the willingness and needs of the editors.

Technological Plan

The e-publishing platform is based on the Open Journals System (OJS), an electronic journal management system that was developed by the Canadian federally funded Public Knowledge Project. As a system that takes care of the whole editorial and publishing process, OJS currently offers assistance at every stage, from the early steps of submission to the final details of publication and access. As mentioned above, the LIC has translated OJS together with a other software from the PKP platform, Open Conference System, into Greek and has distributed it to interested libraries to set up and develop their own journal collections. The LIC administers the system by exercising overall control of accounts, roles and procedures for journal managers and administrators, and by providing end users with the option to freely register to the system and exploit the available personalised services.

Outputs of the Programme

The business plan resulted in two collections, namely 'Pasithee' and 'Dexamene', which are outlined in the following.

Pasithee

'Pasithee' constitutes the core of the e-publishing programme of the LIC as it addresses the improvement of scholarly output through the provision of appropriate technological solutions. Before developing the programme LIC personnel consulted the interested editorial boards to analyse their requirements and hear their views on system development. In a dialectic process the LIC set up the system and editorial processes and instructed the members of the editorial boards.

'Pasithee' currently hosts three journals published by research groups of the University.Electra is an e-journal published by the Centre for the Study of Myth and Religion in Greek and Roman Antiquity of the Department of Literature, University of Patras. Electrapublishes scholarly articles that study the ancient Greek and Roman mythology and religion from literary, historic, anthropologic, archaeological, linguistic or philosophic viewpoints and are written in Greek, English, French and German.

Society and Theory intends to publish surveys and analyses by Greek and foreign scholars from the fields of education, politics, economy, technology, history or philosophy and which, according to the Editorial note, 'have the theoretical and methodological perspective' to fit in this journal. The journal aims to foster the scientific way of research and analysis of social phenomena with articles and announcements of events that will be organised by the editorial board.

Patras Working Papers in Linguistics is a biannual publication of the Centre of Modern Greek Dialects, Department of Philology, University of Patras. The aim of the editorial team is to cover '...any field of Linguistics, synchronic or diachronic, theoretical or applied'. PWPL is a journal that aims to publish special issues on selected interesting topics, with the first issue discussing 'Morphology'. Papers can be submitted in English or Greek.

Dexamene

'Dexamene' is a collection of digital back-files of important Greek printed journals, allowing everyone to access information that would otherwise be difficult to retrieve. 'Dexamene' continuously updates the past content of these journals by means of a moving wall. The moving wall constitutes the main difference between 'Dexamene' as an e-journal archiving system and other collections, like 'Kosmopolis' and 'Pleias' that focus on archiving discontinued journals.

The collection is currently hosting two journals in the field of philosophy and social sciences, Axiologika and Deukalion. Axiologika focuses on the discussion of the social sciences as related to the concept of critique, creating a platform for epistemological discourse. Deukalion is one of the oldest journals for philosophical enquiry and critique in Greece as it was first published in 1970. Its publishing history is divided into three periods of publication. In 'Dexamene' twelve volumes, from 1992 to 2002, are included.

The services offered to 'Dexamene' journals copyright holders are similar to the ones of 'Pasithee'. Once again the library discussed the system and signed a contract with the copyright holders and took the responsibility for administering the system and preserving the content. The LIC digitised the content and imported the relative metadata. Metadata entry was accomplished through an in-house application which was specially developed by the LIC in order to streamline the editorial process.

Horizontal Services: the Directory of Greek Digital Resources

In the past few years Greece has witnessed significant activities with regard to digital collections development. These have led to an increase in Greek resources in digital form, mainly addressing the issues of preservation of historic material and the development of institutional repositories. In this environment the need to coordinate activities in order to reduce the risks of developing duplicate collections and to enhance effective resource management was eminent. Therefore, in the absence of national supervision and administration, the LIC developed a horizontal service, which is called the Directory of Greek Digital

Resources. The Directory is running in a pilot version since February 2008 and aims to become a reference point for useful information about all scholarly and related Greek digital resources available on the web by (a) continuously monitoring and recording all resources with their content policies, access policies, etc., and (b) recording and analysing all basic functional information and parameters for each resource, such as services, technologies, standards, protocols, business models, etc.

The Directory collects data for the following types of resources:

- bibliographical databases: library catalogues, bibliographies, tables of contents;
- digital collections: journals archives, books, texts and articles collections, images and multimedia collections;
- institutional repositories: grey literature, technical reports, pre/post-prints, educational material, archives;
- e-journals;
- experimental and research data collections.

Currently more than 200 resources operated by academic and research institutions, not-for profit organisations and museums, learned societies, private publishers and governmental/state bodies are being listed.

The Directory is already considered to be a valuable tool for the systematic monitoring and analysis of the digital landscape in Greece focussing particularly on the growth of Greek digital content, the adoption of current technologies, models and standards, and on the quality of services to the users. In this way the Directory contributes effectively to the fulfilment of basic objectives, such as recording of resources and initiatives, sharing and dissemination of related information to users and the community, co-ordination and cooperation of institutions' executives and decision makers, follow-up of technical issues, evaluation, etc. (Figure 3).

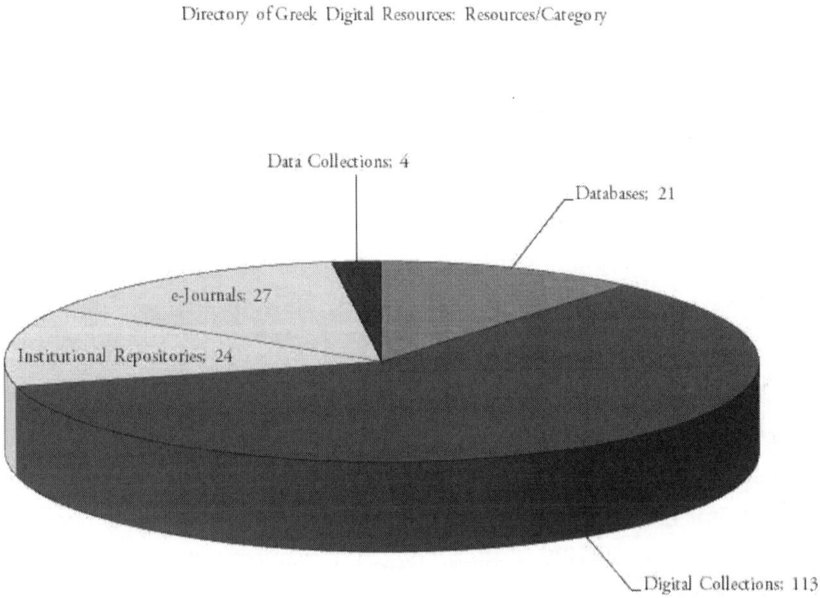

Figure 3. Greek Digital Resources by Category.

DISCUSSION

The LIC's current e-publishing programme is operating in an environment of crisis and transformation of the Greek economy and society. The conditions of this environment seriously affect the economic viability and stability of the programme, but they also challenge the library to respond at the social level by forwarding open solutions to academia and to the public. The programme has its own roots and throughout its planning and implementation was envisaged as a 'feasible' solution. As an organisation that is synonymous with trustworthiness, devoted to preserving forms of intellectual capital, the LIC avoided endangering these integral characteristics by not opting for short-term solutions. Therefore the e-publishing activities are developed as a lightweight programme that exploits every resource of our organisation to support scholars in publishing their research through channels that know no boundaries and require no additional efforts.

The application of the programme is a learning process. The programme currently moves from 'infancy' to 'adolescence', fixing mistakes and following best practices in every single area, from policy making to instruction and assessment of the impact. The cases of 'Dexamene' and 'Pasithee' present two different access models, in accordance with the nature of the published content. It is acknowledged that alternative models exist, which may require the programme and the 'consciousness' to mature. Therefore best practices are constantly explored in order for the LIC to timely adapt to scholars' requirements. Finally, as far as solidifying the outcomes is concerned, the effect of the publications in terms of width and depth of usage must be evaluated in cooperation with the editorial boards. The LIC's utter aim remains to provide useful services for useful resources and therefore automatic gathering of standardized metrics has been introduced.

This programme is an opportunity for the library to claim a leading role inside the institution. Actions must be taken to structurally reshape the landscape The library seeks the appropriate institutional interventions to formally strengthen and further centralise its role. The environment supports the library's endeavours to spearhead an initiative to develop a policy framework between intra-institutional teams and organisations. So far, collaboration is informal and refers to the level of collection development, but with the pool of expertise in several fields from all agents it can extend to areas of policy formation, copyright control and legal aspects, technological advancement and so on.

CONCLUSIONS

The current paper presented the case of an e-publishing and e-archiving programme by an academic library. This programme is a response to several challenges, such as the shrinking of the economic budgets, the growth of Greek digital resources and the spread of Open Access. The strict and narrow margins of economic crisis dictate the strong need to safely navigate the programme by aiming at the horizon of reliability and viability. The programme does not reinvent the library or help the library to identify its role; rather it calls on the library to reinvest knowledge and expertise capitals for the benefit of scholarly communication.

REFERENCES

1. ARL (2009): ARL Statistics 2007–2008. Association of Research Libraries: Washington

2. Brown, L., R. Griffiths, M. Rascoff and K. Guthrie (2007): 'University publishing in a digital age', Journal of Electronic Publishing, 10(3).

3. EKEBI (2009): I biblioparagogi stin Ellada 2008 (text in Greek, title transl.: Book Production in Greece 2008).

4. Henderson, K.S. and S. Bosch (2010): 'Periodicals Price Survey 2010: Seeking the New Normal', Library Journal.

5. Jöttkandt, S. (2010): 'Libraries as Publishers: New Roles for Libraries' (Preprint).

6. Jöttkandt, S., J. Willinsky and S. Kimball (2009): 'The role of libraries in emerging models of scholarly communications', in LIANZA, Christchurch, New Zealand, 13 October, 2009.

7. Rosenblum, B. (2007): 'Building publishing services in the academic library', in Colorado Academic Library Summit, Denver, Colorado, June 1, 2007.

8. SAGE (2009): Meeting the challenges: societies and scholarly communication — Results of a survey to understand the needs of scholarly societies in the 21st century. SAGE, (s.l.).

9. White, S. and C. Creaser (2007): Trends in scholarly journals prices: 2000–2006. Loughborough: LISU

NOTES

[1] http://utsescholarship.lib.uts.edu.au

[2] http://www.ep.liu.se

[3] http://projecteuclid.org

[4] http://www.lib.umich.edu/spo

[5] http://openhumanitiespress.org

[6] http://xantho.lis.upatras.gr/kosmopolis/

[7] http://xantho.lis.upatras.gr/pleias/

[8] http://xantho.lis.upatras.gr/daniilida/

[9] http://xantho.lis.upatras.gr/dexameni/

[10] http://xantho.lis.upatras.gr/pasithee/

[11] http://nemertes.lis.upatras.gr/dspace/

[12] http://www.lis.upatras.gr/Libworld/gr_resources_EL.php

11

CHAPTER

Research on Service Pattern of Storing Local Government Information into University Libraries in Big Data Age

Linlin Zhang

Heilongjiang Bayi Agricultural University Library, Heilongjiang Bayi Agricultural University, Daqing, China

ABSTRACT

Under the big data environment, this paper proposes the system building pattern and effective service strategy of storing government information into university libraries. The local government information can be better co-built and shared by establishing the guarantee institution for government information storage at university library and effectively integrating the government information resources, and providing diversified services of local government information to the public.

Keywords: Government Information, University Library, Storage Institution, Diversified Service, Co-Building and Sharing

1. INTRODUCTION

Although the governments are the largest owner of the information resources in China, these information resources are not effectively exchanged and communicated, most information resources can not be effectively developed and utilized, and information resources are not shared to the society due to no uniform government information storage system. To share the government information resources, we should depend on electronic E-government platform and strengthen resource construction cooperation with the reference collectors in order to ensure smooth sharing of information resources.

"Publication regulation" regulates that the government information is publicized via the public library in China. As a key reference collector, compared to the public library, the university library is suitable for storing the government information due to its strengths in talents, devices and resources. In recent years, partial university libraries have started to serve for local social economy construction and mass and maximize its role as the information and culture communication organization based on their capabilities. e.g. Guangzhou University library cooperates with the governmental authorities to build the Guangzhou governmental consulting hall, which provides governmental information, government archive inquiry, and office guidance [1] . Nankai University and China Mining University integrate the governmental resources and provide reference information services to related industries [2] . These services launch the pattern of publicizing the governmental information resources via the domestic university libraries to some extent. Compared to China, the government information is stored into the university libraries, which becomes the key channel and platform for the public to acquire and utilize the governmental information resource. e.g. the Las Vegas branch library of University of Nevada builds "government information and publication portal website", which updates and arranges the governmental information in continuous change in order to ensure instant and authoritative government information. The reference center website of Michigan University sets Web feed, which can push government information to the public desktop via RSS technology [3] . University of North Texas library cooperates with the America Publication Bureau to develop "government webpage archiving items", which provides permanent public retrieval of withdrawn government websites and electronic publication [4] . Under the big data environment, the university libraries should participate in the local government information service to explore the self-development space, change to the smooth channel for communicating the local government information, and provide more services for co-building and sharing of the local government information resources.

2. TARGET PRINCIPLES AND CONTENT OF GOVERNMENT INFORMATION OF DEPOSIT LIBRARIES IN COLLEGES AND UNIVERSITIES

2.1. Goals of Public Service

The deposit library program of colleges and universities has two main goals for the public service. The public can access to government information resources and it can provide assistance when the public uses government information resources.

2.2. Principles of Public Service

The public service of deposit libraries in colleges and universities needs to insist on two basic principles, including the principle of free access and the principle of comparable treatment. Relevant laws and regulations shall be formulated. The specified deposit libraries in colleges and universities must provide the public with free access to all the government information products. The service of access to government information resources of deposit libraries in colleges and universities shall be equal or more than other library collection services. The reference consulting services of government information resources provided by deposit libraries in colleges and universities to the public must be comparable to the basic library reference consulting service provided by the libraries to the main users.

2.3. Content of Public Service

The key objective of deposit libraries in colleges and universities is to provide the public with free online access of government information resources with public interest and education value. And there is no need to consider the format and media of the information resources. It is ensured that in spite of technological changes in the past or in the future, the users can get assess for the long term. Only the government information purely issued for administrative purpose, and that involving national security secrets, personal privacy and publication in cooperation must be sold, so that the self-sustaining government information can be excluded.

3. STRENGTHS OF STORING GOVERNMENT INFORMATION INTO UNIVERSITY LIBRARIES

With development of information technology, the university libraries will transform to the multi-function and unidirectional reference information center, and extend and expand service functionalities as much as possible to meet demands of more masses. As a traditional technology reference information center, compared to other reference organizations, the university libraries have distinct strengths in storage of the government information.

3.1. Strengths in Hardware Facilities

The university libraries have distinct strengths in network facilities, digital technology and reference carrier. In recent years, the university libraries evolve much in networking, automation and digitalization. Based on the investigation data, the wireless network coverage rate of the university libraries reaches 49.2%. The mean storage capacity of the digital resources is about 44.3 TB. 50% university libraries have 501 - 1000 M backbone network. The Huiwen, Jinpan and ILAS reference systems are extensively used. Many university libraries are united via Internet. They provide better information service experiences to the public via high capacity, high speed and advanced information transmission, so they provide enough hardware conditions for the university libraries to storage the local government information.

3.2. Strength in Information Technology and Talent Supply

Most university libraries have their information management academies, so they have distinct strengths in advanced information technology intervention, and rich software development and information technology application experiences. All university libraries have their perfect reference information organization and classification system, modern information processing technology and information retrieval technology. Librarians can analyze information requirements of the public, proactively provide required information to users and satisfy different demands of users via selection, analysis, arrangement, integration and Push technology. These lay the technology foundation for quick information transmission. Some university libraries establish the resource sharing and cooperation with other university libraries via the network in order to provide deep reference and consulting services to the public.

The university libraries have hybrid librarians in book intelligence specialty, who have higher degrees, professional knowledge and practical experiences, and can arrange and retrieve the information resources. These librarians not only are proficient computer knowledge, but also effectively transmit and manage collected information, provide information retrieval and deep information processing according to different retrieval requirements and retrieval contents of the public, and satisfy information consulting services of the public in different forms and levels.

4. ESTABLISH GUARANTEE INSTITUTION FOR STORING GOVERNMENT INFORMATION INTO THE UNIVERSITY LIBRARIES

4.1. Set government Management Coordination Organization for Effective Operation

Although "public regulation" is issued to make the libraries join in the governmental information publication system. It is difficult to guarantee its status and role on the current phase [5] . To effectively storage government information into libraries, corresponding laws and regulations should be prepared to ensure that the university libraries can restrict governments in information storing. To establish effective cooperation with the governments for a long period, the independent government management coordination organization should be set in the governmental department, so it can regularly ask the electronic and paper government publications and guide storage service of the university libraries in order to effectively storage the government information into the university libraries.

4.2. Establish Take-Over Institution for Storing the Government Information into the University Libraries

The government information should not be only temporarily stored in the university libraries. Although how to storage the government information into the university libraries is in exploration, the university libraries and government organizations should focus on this work and normalize this work, so the laws in organization construction, implementation mode and functionality should be issued, so more masses can know how to acquire the required government information by using the university libraries. The university libraries can better embody the property of the organization and ensure references with their professional strengths.

4.3. Uniformly Prepare Government Information Storing Standard

To better share the govern information resources, the uniform government information description standard should be prepared at the beginning. The metadata from different departments and different government information should be scientifically organized, uniformly indexed, effectively reflected by using the reference classification method, so the unused subject databases can be obtained. The public can quickly inquire the required government information by using the corresponding search engine.

5. EFFECTIVE INTEGRATION OF UNIVERSITY LIBRARIES TO THE GOVERNMENTAL INFORMATION RESOURCES

To storage the government information into the university libraries, the government information center as the leader, universities as dependent party, and government information management coordination organization will compose a digital network library for co-building and sharing of the information resources. This digital network library is the local uniform digital reference information center, which can provide omnidirectional and diversified reference information service to ranks in the society. Three parties should have specific responsibilities and tasks. The local government information center should completely coordinate and lead construction and operation of the whole project. The university libraries should be responsible for development, design and maintenance of the portal website. The government information management coordination organization should be responsible for public advertisement and resource construction. Three parties should select and maintain the resources together and provide effective recommendations and comments to digital network information service.

5.1. Collection Mode and Range of Stored Information in Libraries

The university libraries can be assigned with the storing persons, who will contact the government organizations, and collect, catalog and serve the government information resources. The storing persons can select category information suitable for collection via the collection tools, and form corresponding subject list as the collection list of storing library. The library collection subjects are added or deleted after careful consideration and demonstration. Once the subject items are identified, the related leaders of the government information center should automatically distribute related government information according to the storing subjects of the libraries by using the automated storing distribution system.

To better satisfy diversified information requirements of the social public, the university storing libraries should set collection range of government information in advance. This range can be regularly changed according to the change of the public information demand to ensure that the collected government information resources can satisfy public demands. University libraries should select the subject range, age and format of the government information as wide as possible, so the libraries can provide effective reference consulting and information retrieval service and meet demands of general public. The libraries should set the required items and optional items. The required items should include the very important governmental information and can be classified by the subject list. The optional items include the characteristic government resources, or items strongly recommended by the public. The university libraries can not collect all government information resources. To improve utilization rate of resources and save the librarian time, the university library should cooperate with the government information center in collection, so the storing libraries can provide comprehensive information resource service to the public in the area.

5.2. Information Distribution Management and Storage of Storing Libraries

The university libraries will submit the list of the collected information to the government coordination organization, which will ask the published government information to the government according to the list contents. The government organization can pack the tangible publications and post them to the university libraries, and provide the online link URL of the electronic publications for retrieval, which can support precise retrieval, fuzzy retrieval and resource maintainer retrieval entrance. The university library should establish the correct government information receiving and saving institution to publicize the received electronic government information quickly, and should establish the long-period mechanism for receiving information together with the government. The minimal standard for saving and maintenance should not be inferior to it of other reference type. The references can be saved selectively and permanently. The selective references can be stored for at least 5 years. The discarded government information should be agreed by the government department and be returned. The permanent stored documents should be stored by priority.

6. UNIVERSITY LIBRARIES PROVIDE GOVERNMENTAL INFORMATION TO PUBLIC VIA DIVERSIFIED SERVICE MODE

Now some university libraries have implemented the government information services in China, but many problems occur. The libraries can not closely associate the governments, publicize them in place, and provide narrow acquisition channels to the public, and low-level service. To better serve the local government, the local government should strengthen advertisement, deepen service content, and ask for government support for change.

6.1. Strengthen Internal Training and Learning

The storing university libraries should focus on storing of the government information. The library leaders and general workers should be familiar with storing flow of the government information resources and acquisition mode of resources, so training and learning should be performed inside the libraries prior to external advertisement to ensure that librarians accurately provide timely service consulting service to the public. Training involves use method of automated software, knowing of government laws and regulations, application procedure of general government business, and retrieval method of government websites and database. In addition, training should include access policies, opening time, service standard, device use and collection development of the storing libraries. The university libraries should make the librarians know knowledge on government information via crossing learning and investigation, exchange with the administrative departments and archive library, establish cooperation relation, and provide timely and effective government information service to the public.

6.2. Carry out External Advertisement and Training

To make the public know that the university libraries provide the local government information, the university libraries should advertise the collection service contents the public and make the public know it. e.g. the libraries can advertise via the following means and modes: 1) Post marks of the storing library. The university libraries should post the marks at the remarkable positions, so the public can know that the libraries can provide the governmental information service; 2) Place the advertisement manual. Place the brochure at the remarkable position of the library entrance, so the public can know type and search mode of the stored government information. 3) Establish the website and webpage of stored government information. Embed the government information retrieval portal at the remarkable position on the retrieval pages of the library electronic platform and provide important government information link to the public.

6.3. Carry out Reference Consulting Services via Multiple Modes

To achieve better government information service effect, the libraries can set the public reference consulting center to solve consulting problems on government information in a face-to-face manner, e.g. library collection layout, storage location of government information, use guidance of collected references, consulted issues, and resource commendation, and explain how to acquire the required reference from other libraries via the inter-library borrowing The librarians should further expand the reference consulting service means via post, telephone, email, fax and online service. They can provide different replies according to the type of the public consulting issues to ensure that the public can acquire the useful information quickly.

7. CONCLUSION

In a word, under the big data environment, by combing self-features and strengths, the university libraries should deeply carry out and fulfill the spirit of the central party on public information construction, establish and perfect the guarantee institution for storing the government information to libraries, collect, organize and integrate massive government information in time, make the public effectively acquire the required information in time via one-stop convenient manner, fully release the information energy of the big data, and make the public know the information to most extent.

FUNDING

Daqing Philosophy and Social Science Plan research project "Research on system building and service strategy of storing local government information into Daqing university library" (project number: DSGB2015079).

Heilongjiang Philosophy and Social Science Plan research project "Information service system building of university libraries in big data age" (project number: 14E045).

REFERENCES

1. Chen, J.L. and Wang, J.D. (2009) Analysis on Service Development Pattern and Hot Issue Fields of Local Government Information in China. Library and Intelligence Service, 24, 55-58.
2. Wu, N.Y. (2010) Research on Strategy for University Libraries to Serve the Local Governments. Hebei Science and Technology Map, 3, 32-34.
3. Cheng, M.Q. and Peng, G.L. (2010) Current Conditions and Pattern of Government Information Resource Collection in America Storing Libraries. Library Forum, 2, 29-31.
4. Li, Y.W. and Yu, C.P. (2011) Storing Library Pattern and Application Research. Library Work and Research, 12, 21-23.
5. Yan, L.P., Shen, C. and Zhang, L. (2015) Thinking on Perfecting Local Government Information Publication in Big Data Age. Legality Vision, 1, 38.

Index